Advance Praise for

Solar Water Heating

Bob's book will be required reading for all of our new
installation personnel since it provides a complete and comprehensive history,
guide and information about the type of work we do every day. It is nice
for our new people to get a solid overview of the many types of
technologies and strategies for using solar hot water.

— RICHARD LANE, Managing Partner, Solar Mining Company

Bob Ramlow's *Solar Water Heating* is a practical solar encyclopedia
for anyone interested in installing a solar energy system or in starting a solar
energy company. The folksy writing is clear and the numerous diagrams make
this book a great resource for everyone from the novice just getting started
to the existing solar professional looking to pick up some of
Bob Ramlow's many useful observations from a long career
installing and maintaining solar energy systems.

— DON WICHERT, Director, Renewable Energy Programs,
Wisconsin Energy Conservation Corporation

It's high time someone wrote a comprehensive but accessible
overview of solar heating, which is one of the most cost-effective steps that
we can take to wean from fossil fuels. Bob Ramlow is a rare combination of
plumber, engineer, and public educator who has learned with his own
hands. Whether you wish to do it yourself or simply be an informed
consumer, start with *Solar Water Heating!*

— WINDY DANKOFF, solar industry pioneer and educator since 1975,
and founder of Dankoff Solar Products

Books for Wiser Living from
Mother Earth News

Today, more than ever before, our society is seeking ways to live more conscientiously. To help bring you the very best inspiration and information about greener, more sustainable lifestyles, New Society Publishers has joined forces with *Mother Earth News*. For more than 30 years, *Mother Earth News* has been North America's "Original Guide to Living Wisely," creating books and magazines for people with a passion for self-reliance and a desire to live in harmony with nature. Across the countryside and in our cities, New Society Publishers and *Mother Earth News* are leading the way to a wiser, more sustainable world.

SOLAR
WATER HEATING

SOLAR
WATER HEATING

A COMPREHENSIVE GUIDE TO SOLAR WATER AND SPACE HEATING SYSTEMS

BOB RAMLOW with *BENJAMIN NUSZ*

ILLUSTRATIONS BY BENJAMIN NUSZ

NEW SOCIETY PUBLISHERS

Cataloging in Publication Data
A catalog record for this publication is available from the National Library of Canada.

Cover design by Diane McIntosh.
Cover photo: ©2006, Edward Caldwell. All rights reserved.
All interior illustrations by Benjamin Nusz unless otherwise credited.

Printed in Canada.
First printing, May 2006.

Paperback ISBN 13: 978-0-86571-561-5
Paperback ISBN 10: 0-86571-561-0

Inquiries regarding requests to reprint all or part of *Solar Water Heating* should be addressed to New Society Publishers at the address below.

To order directly from the publishers, please call toll-free (North America) 1-800-567-6772, or order online at www.newsociety.com

Any other inquiries can be directed by mail to:
New Society Publishers
P.O. Box 189, Gabriola Island, BC V0R 1X0, Canada
1-800-567-6772

New Society Publishers' mission is to publish books that contribute in fundamental ways to building an ecologically sustainable and just society, and to do so with the least possible impact on the environment, in a manner that models this vision. We are committed to doing this not just through education, but through action. We are acting on our commitment to the world's remaining ancient forests by phasing out our paper supply from ancient forests worldwide. This book is one step toward ending global deforestation and climate change. It is printed on acid-free paper that is 100% old growth forest-free (100% post-consumer recycled), processed chlorine free, and printed with vegetable-based, low-VOC inks. For further information, or to browse our full list of books and purchase securely, visit our website at: www.newsociety.com

NEW SOCIETY PUBLISHERS   www.newsociety.com

CONTENTS

ACKNOWLEDGMENTS

MOST AUTHORS PAY TRIBUTE TO THEIR FAMILIES, AND FOR GOOD REASON. My family has supported me over the years, has encouraged me to fulfill my desire to pursue a socially responsible career, and have all contributed to this project. I pay special tribute to my wife Marguerite, who is my companion on this path.

It was Dr. George Becker who showed me his solar heating system back in 1971. A light went on in my head that day that has not been extinguished. It was Stephen Morris who encouraged me to start writing about solar and planted the seeds for this book many years ago. And it was the good judgment of my daughter Chamomile to marry Benjamin Nusz, who has diligently worked with me on this project over the last year. His enthusiastic help has made this project happen.

Lastly, I would like to acknowledge the plumbers and solar installers who taught me long ago that there is no substitute for quality. Without the groundwork they laid, I would not be here today, putting the finishing touches on this text.

For the Earth,
Bob Ramlow

INTRODUCTION

EVERY DAY, WE ARE BARRAGED WITH INFORMATION about global warming, global pollution, wars over energy resources, species depletion ... the list goes on and on. Tragically, most people — and certainly most governments — are doing very little about it.

This book tells some of the things we can do. First it talks about using technologies we already have to lower our energy consumption. Then it details renewable energy options for replacing fossil fuels and nuclear energy sources. It shows that as individuals we can change the world!

Most Americans don't think about their energy consumption very much. We don't think about the energy used to create our society or the energy required to produce the products we consume. We get our heating fuels automatically delivered to our homes. Our electrical energy flows to us through silent wires that are everywhere. It's all very easy and painless. Sure, we think about our energy consumption when we pay our bills once a month, or we think about it when we refuel our vehicles at the gas station, but in most cases we just accept these costs as inevitable. Our society uses a lot of energy, we want as much as we can get, and we want it cheap.

The United States consumes more energy per capita than any other country in the world. Other developed countries with similar lifestyles, and with equal Gross National Products, consume up to 50 percent less energy. We obviously have the technology today to use a lot less energy than we do. Some experts say that we can reduce our consumption by up to 75 percent today. They surmise that if we would put our greatest minds to the problem, we could do even more.

Lowering our energy consumption and using renewable energy would accomplish several things. Decreased use of fossil fuels and nuclear energy would have a positive impact on our environment by reducing carbon emissions that contribute to global climate change. It would also reduce the production of pollution that contributes to acid rain, nuclear waste, and other environmental contaminants. A more subtle result of these changes is that power would be transferred from the super-rich to the common people. The people who run the world are the people with the most money. Many of them get their money by selling fossil fuels. These are the people who control the governments and the mass media of the world. By reducing their power we decrease their influence and increase our personal freedom.

It is unlikely that we can look to governments for leadership in the transition to renewable energy that must be made. The changes will come from the bottom up. It is up to us to do what we can.

This book presents detailed information about solar thermal applications for homes and small businesses. Heating water with the sun may seem like a simple endeavor, and it is. Solar water heaters heat water using the energy of the sun. A solar water heating system is composed of a solar collector and a water storage tank. Depending on the type of solar energy system, there may be several other components as well. Most systems use pipes, pumps, system controllers, heat transfer fluids, various valves, heat exchangers, pipe insulation, and mounting hardware. Exact system components will be detailed in the following chapters as we describe the different systems available.

Of all the renewable energy options open to us, the technology of solar water heating is the most mature. It is an industry with proven technologies, established manufacturing facilities, and qualified and experienced technicians. These technologies and applications have been proven over many years and thousands of installations. However, it took time for these technologies to mature. And during that time, we made some mistakes. Some of the technologies or designs that I have included in this book are bad ideas. I have included their descriptions so you can recognize them and know how they work if you are doing repairs. I have also included these descriptions because you may read or hear about a certain design, and you will want to know if it is good or not. Their descriptions should be used only for reference and maintenance.

It does not take a rocket scientist to design, install, or use a solar water heater. While it is simple, certain factors must be taken into consideration to ensure satisfactory reliability and performance. The single most important consideration is your climate. Others include the hardness of your water, hot water demands (the load), aesthetics, specific location factors like trees, shading, and mounting options, availability of components, and price. With over 20 years of experience designing, selling, installing, and servicing solar water heaters, I have found that most systems work great for a very long time. Problems that do occur are generally caused by either poor quality components, the wrong design or components for the climate or specific situation, or poor workmanship. The bottom line is that if you choose the right system for your climate and install it properly, using quality components, you will end up with a renewable energy system that will most likely last you for the rest of your life. It will end up being one of the best investments you ever make.

The next chapter, a short history of solar energy, puts into perspective how we got to where we are today. As you read about the history of solar energy technologies in Chapter 1, you will see that the largest application of these technologies in the world has been in solar water heating.

HISTORY

THE SUN IS THE CENTER of our solar system. The energy it releases warms our planet and powers all life on earth. Through photosynthesis, solar energy is transformed into organic matter — the food that makes our life possible. The fossil fuels we use are actually stored solar energy. Solar energy is also incredibly abundant. Half a day's sunlight falling on the US provides enough energy to run our country for one year.

We often think that modern societies were the first to use solar energy. Not true! Early cave dwellers preferred caves that had openings facing south-easterly. This allowed the morning sun to warm them up without overheating in the warm months. Native Americans in the Southwest oriented their pueblo dwellings so the low winter sun would heat the buildings by direct solar radiation. Cliffs and overhangs blocked the sun during the summer months, helping to keep the dwellings cooler when the sun was high in the sky.

The ancient Greeks, with a climate that was sunny almost year-round, built their houses to take advantage of the sun's rays during the moderately cool winters, and to avoid the sun's heat during the summer. Modern excavations of many classic Greek cities show that individual homes were oriented towards the south and entire cities were planned to allow equal access to the winter sun. It is interesting to note that by 500 B.C., when the Greeks had almost completely deforested their whole country and needed to find a reliable alternative fuel source, they chose solar energy.

The Roman Empire advanced solar technology by adapting home building design to different climates, using clear window coverings such as glass to enhance the effectiveness of solar heating, and expanding solar architecture to include greenhouses and huge public bath houses. Solar architecture became so much a part of Roman life that sun-rights guarantees were eventually enacted into Roman law. This society depleted its forest resource as well.

After the fall of the Roman Empire, the use of glass to enhance solar gain in buildings was mostly forgotten. Interest in passive solar architecture and greenhouses was rekindled during the Renaissance. As technologies advanced, glass manufacturing was revived, resulting in an increased use of glass windows. This also made large greenhouses possible for agricultural purposes as well as for recreation.

In the 1700s, a leading naturalist named Horace de Saussure began to experiment with solar hot boxes. These precursors to today's active solar collectors were simple insulated boxes painted black on the inside and with one side covered with glass. They were very similar to today's solar cookers and, in fact, many early experimenters used their hot boxes to cook with. Many of the solar principles we use today were identified during those early experiments. Unfortunately, these experiments resulted in few successful applications.

During the late 1800s domestic water piped directly into homes became more common. Like today, this water supply was cold. People soon wanted hot running water. At first, all water heaters were either coal- or wood-fired. In 1891 Clarence M. Kemp patented the world's first commercial solar water heater, called the "Climax." It was a black painted water tank mounted in an insulated box with glass on one side.

The Climax was instantly popular in California, where it could be used year-round. Thousands of Climaxes and similar systems were installed in a short time. They all fall into what we now call "batch-type" solar water heaters: the sun heats the water directly in the tank(s) and the hot water is stored right in the collector tank(s).

In 1909 a California engineer named William J. Bailey began selling a new system he called the "Day and Night" solar water heater. It consisted of a solar collector and a separate storage tank mounted above the collector. His tanks were among the first to be insulated for better heat retention and his collectors consisted of a pipe grid attached to a flat plate and enclosed in a compact glazed and insulated enclosure. Cold water dropped down into the collector where it was heated by the sun. As the water was heated, it rose up into the insulated storage tank for later use. Today, we call these heaters "flat plate" collectors.

In 1913 a freak cold snap hit southern California and many Day and Night collectors froze and burst. To eliminate future freezing problems, Bailey installed a coil of pipe within the storage tank to act as a heat exchanger. Then he used an alcohol and water mixture as the antifreeze solution for his heat exchange medium. As the sun warmed the solar fluid, it rose up to the storage tank heat exchanger. As the heat from the solar fluid was transferred to the water in the storage tank, the solar fluid cooled and dropped back to the collectors for further heating. This system is described today as a "closed-loop" solar water heating system.

Between 1920 and 1930, huge deposits of natural gas were found in the Los Angeles area. To capitalize on this new, cheap fuel source, Bailey began to manufacture a thermostatically controlled gas water heater. Sales of his gas water heater took off and sales of solar water heaters plummeted. Gas companies offered generous incentives to hook up to their new gas lines, further hindering sales of solar heaters. Bailey made his last batch of solar water heaters in 1941.

During this same time period, entrepreneurs took the California solar water heater designs to Florida and met with great success. In a building boom between 1935 and 1941, up to 60,000 systems were installed. More than half the population of Miami used solar water heaters by 1941, and 80 percent of the homes built between 1937 and 1941 were solar equipped.

World War II all but halted solar water heater installations. Copper was a major component of solar water heaters and the use of copper was frozen for all non-military use. When the war was over, solar companies came back, but other factors soon led to their decline. Existing solar water heaters were too small to meet the new, increased demand for automatic washing machines, automatic dishwashers, and other similar appliances. In a final blow, electrical rates fell to half their cost before the war, making electric water heating much more affordable. In an aggressive campaign to increase electrical consumption, Florida Power and Light even offered free installation of electric water heaters. By this time, many of the original, aging solar water heaters were experiencing leaking tanks and plugged pipes. Many homeowners found it cheaper to install an inexpensive electric water heater than fix their solar water heating system.

In the United States, the 1950s and 60s were years of unbridled energy consumption. For all but a few, solar energy was a non-issue. This changed with the first Arab oil embargo in 1973, when Americans experienced long lines at gas stations, limited supplies of other oil products like heating fuel, and energy prices that doubled and tripled. President Jimmy Carter helped make energy efficiency and the use of renewable energy a national priority, symbolized by his donning a sweater and installing a solar water heater on the White House roof.

The oil embargo profoundly changed the United States. Coming at the end of the Vietnam War, it added to America's realization of its vulnerability. For the first time since World War II, Americans looked at how they used energy. Consumers began to demand higher energy efficiency standards in everything from homes to automobiles. People also looked to renewable energy sources to replace some of the fossil fuels they were using.

The whole nation took on the challenge of reducing its dependence on oil from the Middle East. Renewable energy sources were rediscovered and new companies sprouted everywhere to fill the growing demand. Government spending on renewable energy research and development increased from around $1 million to over $400 million. While this was a small fraction of the attention and money given to the nuclear industry, it was a dramatic change nonetheless.

During the late 70s and early 80s, installing solar energy systems was seen as patriotic. The federal government, as well as many states, passed legislation encouraging the use of solar energy systems through tax credits. Federal incentives combined with state incentives (where available) often off-set over 50 percent of the cost of many renewable energy systems. A new solar boom began. People looked to wind-powered electric systems, active space heating systems, advanced passive solar heating systems, the newly emerging solar electric systems, and advances in energy saving technolo-gies, as well as the old reliable solar water heaters.

Most of the solar energy compa-nies that sprung up in the 1980s were reliable firms that installed quality sys-tems. Unfortunately, with the headlong plunge into the use of renewables by the general public, a few companies selling inferior products and doing inferior work joined the fray. Some brought products to the market without proper testing. Others just wanted to make a quick buck and didn't care if they were taking advan-tage of well-intentioned consumers. While most renewable energy systems were of good quality, the minority that weren't gave solar a bad name.

The young solar industry was experiencing the typical growing pains that come with most emerging tech-

nologies and took steps to correct the problems. The federal government as well as many state governments also stepped in to ensure higher quality.

This move towards renewable energy did not sit well with those who profited from selling fossil fuels. After the most expensive presidential cam-paign ever, financed in part by oil interests, Ronald Reagan became pres-ident of the United States. His presidency heralded a return to fossil fuels. One of his first acts as president was to remove the solar water heater on the White House that President Carter had installed. Between 1981 and 1986, Reagan effectively gutted the US solar industry. He negotiated a repeal of the tax credit legislation for renewables that were in effect. He reduced funding for renewable energy by 90 percent. He also spearheaded a massive campaign to discredit renew-able energy. The result was a 91 percent drop in the sales of solar hot water collectors between 1984 and 1986. The solar market in North America from the 1980s through the late 1990s was primarily supported by customers who wanted to invest in renewable energy for environmental reasons. For an in-depth analysis of this subject, I invite you to read "Who

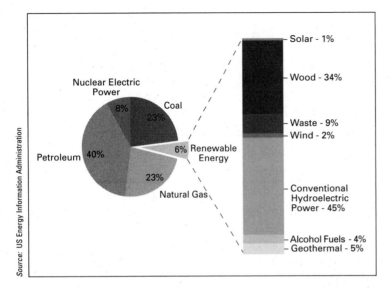

FIGURE 1.1
US energy consumption, 2003

National Product up to one half that of North Americans. They also reduced the energy they need to support a similar lifestyle to half that of the United States. Some countries, like Israel, require solar water heaters be installed on all new dwellings and businesses! Developing nations and Third World countries are also embracing renewable energy technologies over traditional fossil fuel energy sources. While their track record is not perfect, their attitude towards renewable energy is significantly better than that of North Americans.

In Figure 1.1, you can see the breakdown of the total energy consumption for the US in 2003. Renewable energy comprised only six percent! Solar, both thermal and electric, comprised only one percent of the renewable energy share — only 0.06 percent of the total.

Now that we've covered history and the present, we need to look at our future. If what happened in the past carries on through tomorrow, we are headed down a dark and dirty path. One percent of only six percent is not good enough and needs to change now. Let's do all we can to make our children's future brighter.

Owns the Sun?" by Daniel Berman and John O'Connor. Their very thorough and thought-provoking book will change the way you look at energy and politics forever.

While North America was abandoning its use of renewable energy, virtually all other developed nations in the world continued to embrace renewables and energy efficiency. As a result of this continued effort, our European and Asian trading partners reduced their energy consumption as related to Gross

Chapter 2

CONSERVATION AND THE ECONOMICS OF SOLAR WATER HEATING

WHATEVER BROUGHT YOU to this point, whether it was the realization that we are trashing our environment or the simple need to lower your living expenses, now you are here and you want to do something. But what to do first? The answer is simple: start by conserving the energy you use to heat water. Three general principles that are easy to follow will also save you money: reduce losses, increase efficiency, and reduce consumption.

To start, examine your heating system from top to bottom and look for places where heat might leak out. Heat losses in the system end up wast-ing the energy you just used to heat your water. Many can be reduced by just a little bit of cheap insulation. For instance, insulate all of your hot pipes. If you are working on a new construction, insulating the hot pipes is easy. Even if you don't have access to all your pipes, insulating the ones you can get at will make a noticeable difference. You should also insulate your water heater. A tank-type water heater heats a whole batch of water. As this water sits there waiting for use, it slowly cools down. The more you insulate it, the better it will retain its heat. Heat losses can also come from leaks. A faucet that leaks

30 drops of water a minute will waste almost 100 gallons a month. Fix leaky faucets promptly.

Next, try to increase the efficiency of everything in your home that uses hot water, for instance, the washing machine and the dishwasher. Upgrading these appliances to more energy efficient models will significantly reduce the amount of energy consumed. A front-loading washing machine uses half the hot water of a standard top-loading model. This results in around 10–20 gallons of hot water saved in each load you do. You can save thousands of gallons of hot water a year.

Finally, you can conserve energy by simply using less. Much can be done without a significant change in your daily habits. For instance, when washing dishes in the sink by hand, don't let the water run while rinsing. Fill one sink with wash water and the other with rinse water. Soak pots and pans instead of letting the water run while you scrape them clean, and if you are using a dishwasher, only wash full loads. Use cold water with the garbage disposal. Cold water solidifies grease, allowing the disposal to get rid of it more effectively. You can take short showers instead of baths. You should first install a low-flow shower-head. Most standard showerheads use three to four gallons per minute. Even if you take a relatively brief 5 minute shower, you can end up consuming 20 gallons of hot water. Low-flow showerheads will use half of that. A family of four can save well over 1,000 gallons a month. If you are particularly attached to your showerhead, your can install a flow restrictor that will reduce the number of gallons per minute that it uses. For only a couple of dollars, you can reduce your load substantially.

Reducing losses, increasing efficiency, and reducing consumption: these are the first steps. What should be stressed more than quick fixes, though, is the notion of conscious consumption. We have forgotten the financial and environmental costs of hot water. If everyone recognized that whenever they turned on the hot water faucet they were using up energy produced by non-renewable sources, this would reduce energy consumption more than any other measure.

People often say to me, "Bob, I have done a lot of energy conservation and now I am ready to invest in a renewable energy system. What should I do next?"

Today, homeowners and business people can choose from a wide range of renewable energy technologies. Popular options include photovoltaic (solar electric) systems, wind electric systems, and solar water heaters. In almost every case, a solar water heating system is the best place to start. It provides a higher return on your investment than any other renewable energy system. A solar water heater works 12 months a year providing hot water to your home or business with little or no additional costs, thus offsetting your previous bill for heating water with conventional energy sources. Depending on your particular situation, the savings in conventional fuel can pay for the cost of the solar water heating system in as little as three years. Most often the payback is around five to ten years — still a great investment, even without taking into account the ecological benefits of not burning all that fossil fuel.

In fact, since you've already bought this book, I think it's time to let you in on a little secret. Solar water heaters don't cost anything. They're FREE! I know it may sound absurd, but it's true. Now I'm not recommending that you run over to the nearest solar distributor and just take a system. Don't do that. I'm just asking you to take a step back and think

about solar in a different way. With a little change in perspective, you will see that in the end solar water heaters have a net cost of zero dollars.

There are two ways to take this in. The first one is easy: when you install a solar water heater you are increasing your home's value. You gain in equity what you spent on the cost of installation. Solar water heaters typically have a life span of at least 30–40 years. In most cases, the solar collectors will outlast your roof. So if you decide to sell your home, you should get back most of what you paid for the cost of installation. I said this was easily understood, I didn't say it would be convincing. Just because something retains its value over time isn't usually reason enough to go out and buy it.

The second part of this shift in perspective takes a bit more explanation, but I assure you it is even more convincing.

The True Cost of Fossil Fuel

We'll start by comparing solar with the alternatives. Unless you are reading this to find out how to fix your existing system, you probably heat your water with some type of fossil fuel like natural gas, propane, or electricity.

When you purchase fossil fuels you do not pay anywhere near their whole cost. Because our taxes subsidize the oil companies, for instance, the true cost of gas is not reflected in the price we pay at the pump. Let me say it again: oil companies don't pay taxes on all the money they earn so you and I must pay higher taxes to make up for it! It goes without saying that if they paid their fair share of taxes, our tax rates would be lower and the price we pay for fossil fuels would be higher. The same scenario holds true for electricity and all other fossil fuels.

How can this be? First, the fossil fuel companies are among the richest corporations in the world, with tremendous influence in politics. For nearly a century they have manipulated the government into granting them numerous tax breaks and outright payments that are not enjoyed by any other class of corporation. The end result is that they pay little if any taxes themselves, but significantly influence how our tax dollars are spent. They have managed to get the government to pay for lots of expensive research for their industry.

The costs to the environment of using fossil fuels are also hidden. Burning fossil fuels releases carbon into the atmosphere, leading to global climate changes that will disrupt life as we know it on every corner of the Earth. The costs of dealing with these changes will be astronomical, and are directly linked to burning fossil fuels. When we burn fossil fuels, especially coal, chemicals are released into the atmosphere that cause acid rain, polluting our rivers, lakes, and soil. Acid rain kills wildlife, trees, and vegetation, and degrades our buildings, roads, and anything else exposed to it. Although we are already paying some of the costs to fix these problems, we are not paying them all. Eventually, someone will have to pay them.

Then there are health-related costs. Whenever we burn any fossil fuel, pollutants are released into the air that harm our health. Our health insurance costs go up to help pay for the care required by those most affected. Our taxes are increased to help pay for those who cannot afford their own care, and our general health care costs go up for the same reason. Again, we do not pay these costs at the pump or with our utility bill.

Some of our electricity is generated in nuclear power plants. The waste generated by these plants is one of the most toxic substances known to man. We have no clue how to safely dispose of it. We can send men to the moon, but we have not figured out how to deal with these incredibly toxic waste products. Undoubtedly, if we do figure out a way to safely dispose of them, it will be incredibly expensive. This cost is not included when we pay our electric bill. I also invite you homeowners and renters to read the fine print of your insurance policies. Note that if there is ever an accident involving nuclear fuel or waste, your insurance policy does not cover it. Ask the people living around the Three Mile Island nuclear power plant, whose lives were devastated by the nuclear accident there, how they feel. I will guarantee you that they are not happy. Many lost everything.

Assigning a true cost to the use of fossil fuels relates directly to solar water heaters and any other renewable energy system. Admittedly, it costs money to invest in renewable energy equipment. Often, people will look at that cost and say that it is just too much more than using fossil fuels. When you get to this point, please remember the above discussion. How much higher should the costs of fossil fuels really be? Twice as expensive? Three times? Four times? Experts who

have spent considerable time researching this issue have calculated that these costs are five times more than the bill we pay at the pump or meter.

While these true costs are not reflected in our bills, we do see that the price we pay tends to increase every year. In order to understand where fossil fuel prices are heading, we first have to understand what has happened in the past. Having a handle on energy price inflation is basic to understanding the economic impacts of investing in solar thermal energy systems.

A Brief History of Fossil Fuels

Soon after oil was first discovered in the mid-1800s there was a glut, keeping its price very low. With the advent of mass-produced automobiles, there was a steady demand for oil and prices became stable, rising at a rate at or slightly above inflation. During World War II, fossil fuels were diverted to the war effort, so they became hard to get and more expensive. After the war, fuel again became plentiful and relatively inexpensive. During the postwar period, per capita consumption of fossil fuels skyrocketed.

It is important to note that until the 1970s, almost all the oil used in the United States was produced here. In 1970 the United States reached peak oil production while demand continued to escalate. About 1996, imported oil overtook domestically produced oil for use in US consumption.

THE OIL EMBARGO

Beginning in 1973, political factors caused a shortage of oil in the worldwide market, followed by a global recession. The shortage continued through the early 1980s. The OPEC oil embargo made people think about how they used energy, and energy conservation became common practice. For the first time, people began to talk about running out of oil on a large scale. In fact, though, during this period there was plenty of oil available and in the ground. The oil spigots could have been opened at any time and the crisis would have been over in a day. In fact, this is essentially what happened in the mid-1980s.

PEAK OIL AND NATURAL GAS

Today, we are facing an oil shortage much different than that of the 70s and 80s. We are entering the era of peak oil. There are many good books on this subject, like *The Party's Over* by Richard Heinberg, so I will not go into

a lot of detail here. But essentially, today we are at a turning point in the history of modern civilization because the production of oil is at its peak; it will never grow larger, as it has in the past. At the same time, worldwide demand for oil is growing faster than at any time in history. As a result, the price of oil will continue to rise while the supply will decrease. It is important to note that the earth is not running out of oil. About one-half of all the oil there ever was is still left in the ground. The reality is that we have reached peak oil *production,* while demand for oil continues to rise at record levels. For some time there will still be oil to be had, but producing it will become increasingly more expensive.

You are probably wondering why I am discussing oil at such length. While oil is rarely used to heat water, the price of oil affects the price of all other forms of energy. When it goes up, they do too. This is especially true for electricity. It takes large amounts of oil to mine coal, the basic feedstock of most of our electrical generating capacity in the US. It takes oil to mine and process uranium to feed our nuclear power plants. Oil is used in the natural gas exploration and distribution industry. Our society is completely and utterly dependent on a constant flow of cheap oil. As we enter the peak oil era, its price will continue to rise, with no end in sight.

The same holds true for natural gas, which *is* used to heat water. We have now reached peak natural gas production in North America, where all the natural gas used in the United States is produced. We will never be able to produce more natural gas than today, even if demand rises (which it is doing). When the United States reached peak oil production in 1970, this was a significant milestone, but it was not devastating to our economy or culture. The oil companies could inexpensively import oil from other parts of the world. All they had to do was to build inexpensive oil tankers to ship foreign oil to the US. This is not the case for natural gas. It takes very sophisticated and expensive ships to import natural gas from foreign sources. Also, expensive and sophisticated terminals must be built at both the shipping and receiving ports. This infrastructure is not in place and it will take many years and a substantial investment to create it. This will significantly affect the price of natural gas in the future. The bottom line is that the cost of natural gas will also continue to rise, with no end in sight.

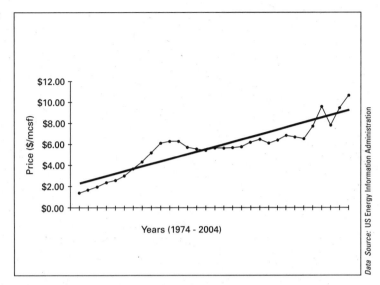

Data Source: US Energy Information Administration

FIGURE 2.1

US residential natural gas price trends

Life Cycle Costing

People often ask, "Why would I consider purchasing a solar water heater that costs several thousand dollars when I can purchase a gas or electric water heater for only several hundred dollars?" The answer lies in the fact that they do not think about *life cycle costing*. Life cycle costing adds the original cost of a piece of equipment to its operating cost over the equipment's lifetime, or at least over a certain amount of time. Using an analysis like life cycle costing gives an accurate analysis of the real overall cost of a purchase and allows you to make accurate and informed comparisons.

This is why it is important to figure out the *energy inflation rate*. We know, for example, that natural gas prices will increase rapidly. The question remains: by how much? The energy inflation rate is simply the percentage more you have to pay each year for the same amount of energy. The average energy inflation rate over the past 30 years for natural gas in the residential sector was 7.54 percent.

Figure 2.1 graphs the price of natural gas over that time period. In the late 70s and early 80s you can see the effects of the oil embargo and how the surplus lead to a rate decrease. You can also see what has been happening lately — the sharpest rate increases since the oil embargo, with inflation rates that surpass those of that era. Over the past 5 years natural gas has increased at an average rate of 11.10 percent, and this is just the start. The rate increases of the last couple years have been over 20 percent.

You can't just use the cost of energy during the first year for each additional year because we know that the cost is rising at such a sharp rate. The lowest energy inflation rate that can be used is the historical 30-year average of around 7.5 percent. However, using this rate does not take into account the fact that dwindling supplies and increased demand will have a strong impact on rate increases. Conversely, using an inflation rate in the 20 percent range like we have seen in the last couple of years may result in unrealistic estimates. We made that mistake before, during the oil embargo. Nevertheless, it is best to err on the side of caution and try to be conservative in your estimates without being unrealistic. When I estimate the life cycle costing, I use ten percent as the energy inflation rate. I think that this is the lowest it will realistically be over the next 30 years. Most likely we will see the next 30-year average in the 15–20 percent range.

In Figure 2.2 you can see an example of life cycle costing. This example compares both an electric water heater and a natural gas water heater to a solar water heater. All three systems are producing the exact same amount of hot water. But solar water heaters have NO OPERATING COSTS. Like any other piece of mechanical equipment, they do require some maintenance, but this only amounts to around $2.00 per month.

As you can see in the table, viewing the systems in the long term makes for a more fair comparison. The cost of the solar water heater is equal to the operating cost of the electric water heater after only 8.5 years and the natural gas water heater after 14.5 years. This number is commonly referred to as the pay-off date, since you would have paid for the system with the money saved from not having to purchase energy from the utility. However, I want to stress that this is a misstatement. As I said before, a solar water heater is paid off the second you install it because of what you have gained in equity.

Nevertheless, the notion of a pay-off date is still a good way to think about the cost of a solar water heater. For instance, imagine yourself nine years from now. You will have taken just as many showers, washed just as many loads of clothes, essentially used the same amount of hot water, but you had a choice whether to heat this water with polluting coal or with clean energy from the sun. You would have paid the same amount over that time period. In other words, a solar water heater will not cost you a penny more. The bottom line is that over a nine-year period the two systems cost

		Electric Water Heater	Natural Gas Water Heater	Solar Water Heater
Energy Produced		4,300 kWh	200 therms	200 therms or 4300 kWh
Cost per Unit		$0.10	$1.00	$0.00
Energy Inflation Rate		10%	10%	
Installed Cost		$1,500.00	$1,500.00	$7,000.00
Maintenance		$0.00	$0.00	$2.00/month
Cost to Operate	1st year	$430.00	$200.00	$0.00
	2nd year	$473.00	$220.00	$0.00
	3rd year	$520.30	$242.00	$0.00
	4th year	$572.33	$266.20	$0.00
	5th year	$629.56	$292.82	$0.00
	6th year	$692.52	$322.10	$0.00
	7th year	$761.77	$354.31	$0.00
	8th year	$837.95	$389.74	$0.00
	9th year	$921.74	$428.72	$0.00
	10th year	$1,013.92	$471.59	$0.00
Total Cost of Energy Consumed		$6,853.09	$3,187.48	$0.00
Life Cycle Cost				
	After Ten Years	$8,353.09	$4,687.48	$7,240.00
	After 15 Years	$15,162.17	$7,854.50	$7,360.00
	After 20 Years	$26,128.25	$12,955.00	$7,480.00
	After 30 Years	$72,232.43	$34,598.80	$7,720.00
	After 40 Years	$191,814.80	$90,018.51	$7,960.00

FIGURE 2.2
Life cycle costing comparison

essentially the same to purchase and operate. It is in this sense that solar water heaters are free. Taking cold showers and not washing your dishes and clothes isn't an alternative. You are going to use hot water in any case, so why not make the responsible choice? Choose solar.

After the pay-off date, the solar water heater will produce free energy for the duration of the system. It truly is free because you have already offset the cost of installation with money saved. With scheduled maintenance, the solar water heater has at least a 40-year life expectancy, so over that time frame the savings from the solar water heating system will be around $190,000.00! That is a lot of money to save from a one-time investment of $7,000. Because of the cumulative effect of energy inflation, the savings add up quickly, totaling quite the impressive figure. Obviously, it makes economic sense to choose the solar water heater over the electric water heater.

The life cycling cost example in Figure 2.2 has been simplified for the sake of argument and fails to take a number of variables into account. The analysis does not include the cost reduction resulting from subsidies for installing renewable energy. For instance, at the time of writing, a federal tax credit for solar water heaters pays for 30 percent of the cost of an installed system, up to $2,000. There are also many states and utilities that

offer rebates for solar. If you are lucky enough to live in one of these states, I urge you to take advantage of these rebates now. Any reduction in the cost of the system will also drive down the pay-off date. Where I live, in Wisconsin, many systems installed today are being paid off in only five years.

Deciding on what fuel inflation rate to use will also affect a comparison like this. If we had a crystal ball that would accurately predict future fuel prices, we would obviously have a more accurate prediction of the actual outcome of the comparison. If energy inflation rates rise at more than just the ten percent used above, the pay-off date will be reduced. If the last couple of years are indicative of the future energy inflation rate and it turns out to be 20 percent, the system would be paid off in 6.5 years. Conversely, if the inflation rate dropped down to the historical average of 7.5 percent, it would take 9.5 years to pay off the system.

Nor does the example include the cost of borrowing if the solar energy system or the electric or natural gas water heater has to be financed. Many of us don't have $7,000 readily available to spend on a system that heats water. That means you are off to the bank to take out a loan. The added interest will add some time to the pay-off date, but only about a year.

And remember, the price of energy shown on your monthly utility bill overlooks the real cost of burning fossil fuels. Each and every year it is in use, the solar water heater used in the comparison above would eliminate around two tons of greenhouse gases for an electric water heater and one ton of greenhouse gases for a natural gas water heater. If the ozone had a dollar value, what would it be?

CASH-FLOW ANALYSIS

Another way to look at the economics of solar water heaters is to look at a cash-flow analysis. This looks at the impact an investment will make on your cash flow. We all have some method of making money and we get a certain amount of money per week as income. Then we spend this income to get the things we need and want. These are our expenses. A sound cash flow is a balance where our expenses do not exceed our income.

A solar water heating investment is different from most investments we make because the value of the free energy a solar water heater harvests reduces a bill you would otherwise pay each month. If you heat your water with fossil fuel, you have a hot water bill each month that is part of your normal cash flow. When you install a solar water heater, your hot water bill is reduced. The savings gained from the solar water heater pays for the solar investment.

Figure 2.3 gives an example of a cash-flow analysis for a solar water heater. I use the same starting costs that were used in the life cycle analysis. The original cost of the solar water heating system is still $7,000. Now assume that you need to borrow all the money to pay for the solar water heating system at six percent interest and make equal monthly payments for ten years. As you can see, the monthly loan payments initially exceed what is saved from not having to purchase electricity. However, over time the two columns level out, making much more comparable figures. Eventually the monthly loan payments are less than the monthly utility bill, meaning that you will actually have more money in your pocket from month to month. After ten years, when the loan has been completely paid off, your cash flow per month is greatly increased and will continue to increase as energy prices rise.

	Electric Water Heater Monthly Savings	Solar Water Heater Monthly Payment	Cash Flow Impact per Month
Monthly Bill			
1st year	$35.83	$77.71	-$41.88
2nd year	$39.42	$77.71	-$38.29
3rd year	$43.36	$77.71	-$34.35
4th year	$47.69	$77.71	-$30.02
5th year	$52.46	$77.71	-$25.25
6th year	$57.71	$77.71	-$20.00
7th year	$63.48	$77.71	-$14.23
8th year	$69.83	$77.71	-$7.88
9th year	$76.81	$77.71	-$0.90
10th year	$84.49	$77.71	$6.78
11th year	$92.94	$0.00	$92.94

FIGURE 2.3

Cash flow analysis solar vs. electric

FIGURE 2.4

Cash flow analysis — with rebate

	Electric Water Heater Monthly Savings	Solar Water Heater Monthly Payment	Cash Flow Impact per Month
Monthly Bill			
1st year	$35.83	$55.51	-$19.68
2nd year	$39.42	$55.51	-$16.09
3rd year	$43.36	$55.51	-$12.15
4th year	$47.69	$55.51	-$7.82
5th year	$52.46	$55.51	-$3.05
6th year	$57.71	$55.51	$2.20
7th year	$63.48	$55.51	$7.97
8th year	$69.83	$55.51	$14.32
9th year	$76.81	$55.51	$21.30
10th year	$84.49	$55.51	$28.98
11th year	$92.94	$0.00	$92.94

Now examine Figure 2.4. This table assumes that the solar water heater was eligible for some type of subsidy. If you purchased the system while the federal tax credit was available, a $7,000 system would only cost $5,000, and the monthly payments would decrease substantially. Essentially, you could have a solar water heater for only an additional $20 a month, and this amount decreases over time. This example shows a small negative impact on your cash flow for the first five years, and a positive cash flow impact thereafter. Of course, after the loan is paid off in ten years, all the savings contribute to increasing positive cash flows. Since the system is expected to last around 40 years, you can plan on seeing many years where you will have more money to spend on a monthly basis. When you finance the system you are essentially locking in your monthly payments. You know what you will have to pay each month and will not be affected by the continually rising cost of energy. The point of this analysis is to show you that you should not get hung up on the upfront cost of a solar heating system because the investment does not significantly impact your cash flow.

At the start of this chapter I told you that I would demonstrate how you can get a solar water heater for free. The point of the life cycle costing and cash-flow examples is to show that no matter how you look at it, a solar water heater will not cost you any more than its alternative. All you have to do is install the system and you can start saving today.

Chapter 3

TYPES OF SOLAR COLLECTORS

ICS Collectors

ICS STANDS FOR Integral Collector Storage. In an ICS unit, the hot water storage tank is the solar absorber. The tank or tanks are mounted in an insulated box with glazing on one side and are painted black or are coated with a selective surface. The sun shines through the glazing and hits the black tank, warming the water inside the tank. Some models feature a single large tank (30–50 gallons) while others feature a number of metal tubes plumbed in series (30–50 gallon total capacity). The single tanks are typically made of steel, while the tubes are typically made of copper. These collectors weight 275 to 450 pounds when full, so wherever they are mounted, the structure has to be strong enough to carry this significant weight.

ICS collectors are widely used around the world in climates that never experience freezing conditions. They work great, given their climatic restrictions. They are a direct type of system as the water you use is actually heated in the collector. They do not suffer from hard water problems nearly as much as do flooded collectors (collectors with hot water in them at all times), and their simplicity

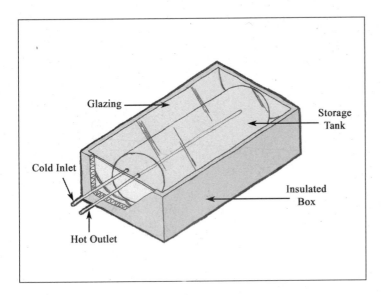

FIGURE 3.1
Tank-type ICS collector

FIGURE 3.2
Tube-type ICS collector

makes them a very affordable choice. ICS collectors are also used in seasonal applications like campgrounds and summer homes where they are only used during the warm months of the year and are drained before freezing conditions occur. I have sold many of these collectors with excellent results.

As mentioned above, these collectors are heavy, so if they will be mounted on a roof, make sure that roof is strong enough to hold their weight. You may have to reinforce an existing roof to make it acceptable. You should always mount these collectors tilted so they will properly drain.

The tube type of ICS collector will outperform the tank type because more surface area is exposed to the sun. Another advantage of the tube type is that their profile is much smaller, which affects their aesthetics. On the other hand, tube-type collectors cool off more quickly at night because of their larger surface area. On cool nights, the water stored in these collectors will cool off, so they lose efficiency. You can maximize this kind of system's efficiency by using as much hot water as you can during the day and early evening hours.

Another kind of collector/system is often classified as an ICS system.

These use a flat plate or evacuated tube collector mounted directly to a storage tank that is located directly above and attached to the collector. These systems use a heat transfer fluid that flows through the collector. When heated, it rises to a heat exchanger located in the storage tank. Heat tubes in the collector can also be used to transfer the heat from the collector to the storage tank. These systems, the majority of which use flat plate collectors, are very popular in southern Europe and areas of Australia where freezing conditions never occur. They work very well because the surface area of the collector is large and because the flat plate absorber is very efficient. These systems do not lose their heat as quickly during the evening hours because the storage tank is well insulated.

All ICS systems are plumbed in series with the back-up water heater and act as a pre-heater. Very often, they can provide 100 percent of the daily domestic hot water when installed in hot, sunny climates.

Flat Plate Collectors

Flat plate collectors are the most widely used kind of collector in the world for domestic solar water heating

FIGURE 3.3

Thermosiphon-type ICS collector

and solar space heating applications. These collectors have an operating range from well below 0°F to around 180°F, which is precisely the operating range required for these applications. They are durable and effective. Flat plate collectors have a distinct advantage over other types in that they shed snow very well when installed in climates that experience significant snowfall. They are the standard to which all other kinds of collectors are compared. I only use flat plate collectors in systems I design or install and I highly recommend them.

Flat plate collectors are rectangular shallow boxes that typically are 4' wide x 8' long and 4" to 6" deep, but also come in 4' x 10' and 3' x 8' sizes. These collectors are made with a strong frame, a glazing is fastened to the front of the collector, and a solid back is provided. An absorber plate lies just beneath the glazing. This absorber plate has manifolds that run across the top and bottom of the collector, just inside the frame. These manifolds are usually one-inch diameter copper pipe and extend out both sides of the collector through large rubber grommets. The manifold ends that extend out the sides of the collector are often called nipples. These collectors, called internally manifolded collectors, can be easily ganged together to make large arrays. Smaller riser tubes, typically half-inch copper pipes, run vertically and are welded to the manifolds above and below and are spaced three inches to six inches apart (the closer the better). Attaching a flat copper fin to each riser completes the absorber plate. The fin must make intimate contact with the riser tube in order to facilitate effective heat transfer from the fin to the tube. Soldering or welding the fins to the tubes makes the best connection. The fins are also usually made of copper and are either plated with a selective surface or painted to maximize solar absorptivity. They are also usually dimpled or corrugated to increase absorptivity. The absorber plates are not attached to the frame, they just sit inside it and can expand or contract as they are heated or cooled without being restricted by the frame. The fins and tubes must be of the same metal to reduce the chance of corrosion. There is one absorber on the market that has copper waterways with aluminum fins. They get away with this by laminating the metals

FIGURE 3.4

Flat plate collector

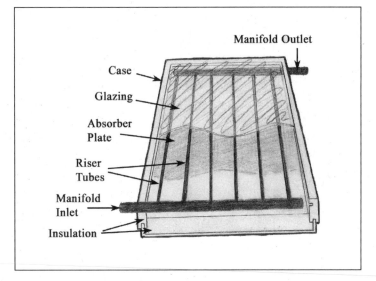

Case

Glazing

Absorber Plate

Riser Tubes

Manifold Inlet

Insulation

Manifold Outlet

together in such a way that galvanic reactions do not take place.

Both test results and experience have proven that one of the best absorber coatings is a selective surface made of black nickel. In order to get black nickel to adhere to copper, the copper must first be plated with chrome and then the nickel is plated to the chrome. Some recent developments in new selective surfaces show good promise. Time will tell if these new surfaces will stand the test of time.

It is very important that a flat plate collector have a strong frame. The collector mounting hardware is fastened to the frame, and strength is very important because the collectors must be able to withstand very high wind conditions without breaking apart. These frames are almost exclusively made of extruded aluminum, although some are made of rolled aluminum. I only choose collectors that have heavy, thick, extruded aluminum frames. The extruded aluminum frames have channels or flanges built into them that the mounting hardware fastens to. Because these flanges go completely around the collector, great flexibility in mounting options is available.

Another important component to look at when considering a collector is the kind of fasteners used to assemble the collector. All fasteners should be made of stainless steel. It is critical to always use compatible metals where they are attached to each other. Aluminum and stainless steel are compatible, while aluminum and plain or galvanized steel are not. This must be applied not only to the construction of the collector, but also to the mounting hardware. Because each manufacturer makes their own mounting hardware, and because each collector is tested with its specific hardware, you should always purchase your mounting hardware to match your collectors.

All kinds of plastics have been used as glazing material for collectors, but they have all failed under direct, constant exposure to the sun. Only low-iron tempered glass has stood the test of time. This glass is usually patterned on the outside to reduce glare and reflection and to increase absorptivity. A rubber gasket is fitted to the edges of the glass plate to both protect the edge as well as to create a good seal where it sits against the collector frame. Note that if you ever have to take the glazing off a collector, the edge of the tempered glass is very fragile. If you even tap the edge or side of a tempered glass pane, it can literally

Outlet

Path of Solar
Fluid

Heat Exchanger
(Condenser)

Inlet

Insulated Box

Absorber Plates

Heat Pipe

Vapor and Condensed
Liquid within Heat Pipe

Evacuated Glass Tubes

FIGURE 3.5
Evacuated tube collector

the tube. During the manufacturing process, a vacuum is created inside the glass tube. The absence of air in the tube creates excellent insulation, allowing higher temperatures to be achieved at the absorber plate. Their similarity, however, ends there.

Some collectors have a riser tube fastened to the absorber plate; it sticks out of each end of the tube and is attached to a manifold, much like a flat plate collector absorber. The solar fluid circulates through each tube and is heated. Others use a hollow pipe that is attached to the absorber plate; it is closed at one end and exits the glass tube on the top. A special liquid is placed inside the pipe and when heated, the liquid evaporates. The hot vapor then rises to a heat exchanger in a manifold located along the top of the tubes. Solar fluid is heated in the exchanger and is circulated throughout the system. As the solar fluid cools the vapor, it condenses and drops back down into the pipe. These liquid/vapor tubes often require a valve on each pipe. Still others use a solid metal rod attached to the absorber; the rod sticks out the top end of the glass tube. When the sun shines on the absorber, the rod gets hot. The heat is then conducted through the solid metal rod.

explode, so be very careful and always wear safety glasses and gloves when handling glass.

Evacuated Tube Collectors

While flat plate collectors are all essentially made the same way and perform the same way from one brand to another, evacuated tube collectors vary widely in their construction and operation. Evacuated tube collectors are constructed of a number of glass tubes. Each tube is made of annealed glass and has an absorber plate within

The end of the rod protruding out the end of the glass tube is inserted into a manifold. When solar fluid is circulated through the manifold, it picks up heat from the rod ends.

There are potential problems with evacuated tube collectors that you should consider. First, these collectors get hotter than flat plate collectors and can generate temperatures above the boiling point of water. This can cause significant problems in a solar water heating or solar space heating system. It is therefore critically important to always make sure there is an adequate load on the system to keep the temperatures below 210°F. Second, the tubes are fragile. They are constructed using annealed glass, which is much more delicate than tempered glass. Care must be taken when transporting and handling the glass tubes. Another concern is that the collector performance relies on the vacuum inside the tubes. Because a rod or pipe exits the tube on one end or both, a seal must be maintained at this junction. Most manufacturers of evacuated tube collectors only guarantee this seal for 10 or 15 years, if at all. If the seal is broken, the performance is no better than a flat plate collector's, and probably worse. Another big issue with evacuated tube collectors is that they do not shed snow. Because the evacuated tubes are such good insulators, little heat escapes them and the snow that accumulates on the tubes can stick for a long time. Their surface is also irregular, so snow packs between the tubes as well. I have seen many instances where roof-mounted evacuated tube collector arrays got packed with snow in the early winter and stayed that way till spring, which rendered them completely useless for a good portion of the year. The glass is too fragile to scrape accumulated snow off as well.

Comparing Flat Plate and Evacuated Tube Collectors

Since their conception, much has been said about the efficiency of evacuated tube collectors. They are commonly heralded around the solar industry as a more efficient collector. If you were just to look at the collector designs, that's an easy assumption to make. Normally, if you reduce heat losses, in this case by the vacuum of the tubes, you would increase efficiency. So it may sound counterintuitive if I tell you that flat plate collectors are actually more efficient for all residential and most commercial applications even

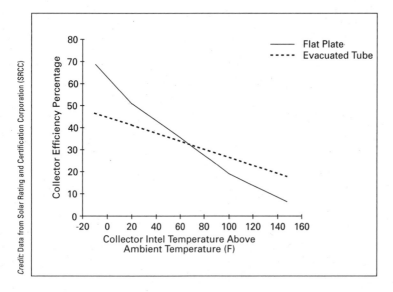

FIGURE 3.6

Mean collector efficiency ratings

though they are not as well-insulated. It takes more than a clever design to produce hot water. It takes a system that is appropriately designed to do what you want it to do.

The measure of a collector's efficiency should really be how they perform when they are put to use. The best way to measure that is through an independent testing organization. One of the good things that came out of the tax credit era was the Solar Rating and Certification Corporation (SRCC). The SRCC rates and certifies many of the collectors on the market today. It is the most common and reliable source here in the United States for independent information about solar collectors. I would strongly suggest buying collectors that they have certified. Not only does the SRCC test for collector performance and efficiency, but also for durability and reliability. Both are critical for determining the value of a collector. The results are free to the public and can easily be accessed online at <www.solar-rating.org>. Using the SRCC gives us good, solid standardized data for comparing collector performance.

When rating a collector, the SRCC will test the amount of heat, in Btu, that it will produce based on a certain amount of radiation that shines on the collector. They usually do this with big lights to ensure consistency between tests. Because the sun doesn't shine every day, they test under three conditions: clear day (2,000 Btu/ft²/day), mildly cloudy (1,500 Btu/ft²/day), cloudy day (1,000 Btu/ft²/day). They mimic how the amount of sun will vary depending on location and climate. As a second variable, they will alter the temperature at the site. This is actually the difference of the temperature of the fluid going into the collector (inlet temperature) and the temperature outside (ambient temperature). Figure 3.6 graphs the ability of each type of collector to convert sunlight into usable Btu for all of the temperature variables and averaging three sun conditions as a measure of overall performance.

As you can see, the flat plate collectors are more efficient until the inlet temperature is around 70°F more than the ambient temperature. Let's consider an example. Let's say that it is 30°F outside and you are putting fluid into your collector that is the same temperature as your ground water, let's say 55°F. In this condition, you would look to the point at 25°F on the graph. Flat plate collectors are around 54 percent efficient, while evacuated tube collectors are only 38 percent. That's quite a difference in performance, even on a relatively cold day. If you are properly dumping the heat, the inlet temperature on most residential applications is usually at most 100°–110°F. However, at that point you will not need much more to reach your desired temperature. The truth of the matter, as the data shows, is that for most conditions, flat plate collectors will outperform evacuated tubes. Now, if you needed really high temperatures, say over 160°F, then evacuated tubes

might be the right collector for the job. But until that point, they simply won't perform as well.

The second claim made for evacuated tube collectors is that they are better collectors during cloudy conditions. Figure 3.7 graphs the efficiency ratings for all three SRCC conditions, including cloudy, low sun weather. As you can see, the point where the collectors' efficiency ratings cross is less than the average, signaling an increased efficiency. However, they are still not more efficient than flat plate collectors in most temperatures.

FIGURE 3.7

Collector efficiency ratings overall all conditions

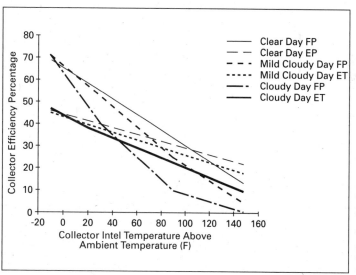

Additionally, you need to consider the value of being better at harvesting a decreased resource. If there isn't much solar radiation to gather in the first place, being slightly better doesn't amount to a whole lot of Btu. More of a little bit is still only a little bit.

The efficiency of the collector should be a major factor when selecting one for your application. Probably equally important is cost. At today's prices, a good flat plate collector will cost around $25 per square foot while an evacuated tube collector will be around $75 per square foot. Evacuated

tubes are three times as expensive. If they were three times as efficient or produced three times as much hot water the additional cost might be justified. But I can't see why anyone would want to pay more for a less efficient collector. Flat plate collectors usually last 30–40 years if well maintained but evacuated tube collectors consistently lose their vacuum after only around 15 years. Remember, too, the earlier figures for pay-off dates: the more you pay for your system, the longer it takes to pay it off. By the time you have your evacuated tube collector system paid for, it will be time to buy some new collectors. I would hope that by then you would have learned your lesson and would replace them with a flat plate collector array.

Evacuated tubes definitely have a secure place in the solar industry in applications where high temperatures are needed. However, their shortened maintenance schedule and decreased overall lifespan in residential applications are a significant setback and shouldn't be overlooked.

Do evacuated tube collectors heat water? Yes they do. But are they the best choice for residential domestic hot water and space heating? I would

FIGURE 3.8

Concentrating collector

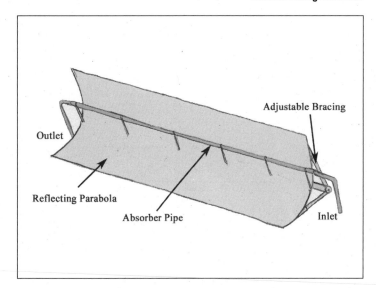

have to say that based on the efficiency ratings and the temperature range of use, no. I didn't put them on my home and I didn't put them on my mother's.

Concentrating Collectors

A concentrating collector utilizes a reflective parabolic-shaped surface to reflect and concentrate the sun's energy to a focal point where the absorber is located. To work effectively, the reflectors must track the sun. These collectors can achieve very high temperatures because the diffuse solar resource is concentrated on a small area. In fact, the hottest temperatures ever measured on the earth's surface have been at the focal point of a massive concentrating solar collector. Concentrating collectors have been used to make steam that spins an electric generator in a solar power station. This is sort of like starting a fire with a magnifying glass on a sunny day.

While there have been attempts to use concentrating collectors in domestic water heating systems, no successful or durable products have been developed. Problems have been encountered with the tracking mechanisms, the precision needed in the mechanisms, and the durability of the reflectors and linkages. Another big

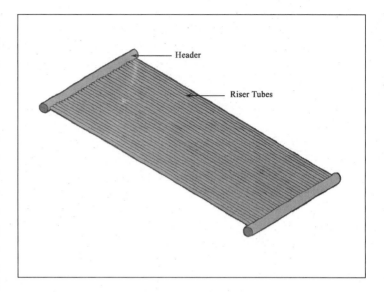

FIGURE 3.9
Pool collector

problem is that they just make the water too hot on a regular basis.

Pool Collectors

The single largest application of active solar heating systems is in heating swimming pools. Special collectors have been developed for heating seasonal swimming pools: they are unglazed and made of a special copolymer plastic. The collectors don't have to be glazed because they are only used when it is warm outside. These collectors cannot withstand freezing conditions.

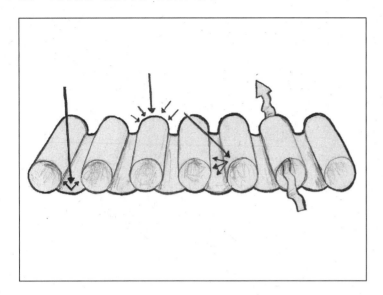

FIGURE 3.10
Web and tube configuration

The absorber plate in copolymer collectors is an extruded mat that has many waterways or risers that are set close together. Many designs have been produced, but the design that has proven most successful is a web and tube configuration. With this design the tubes are separated by a small web, so they are not touching one another. Much like copper absorber plates, these collectors have a header running along the top and bottom of the absorber. The best designs, like web and tube absorbers, are made so that each riser is individually attached to the header, which allows the least backpressure within the collector and allows individual connections to be repaired, if needed. Because the risers are situated close together, this type of collector has a large wetted surface, adding to its efficiency. In fact, these collectors can outperform glazed collectors in pool heating applications.

Solar pool heating systems that use unglazed plastic collectors are direct systems where the pool water is circulated through the collectors. This is the most efficient configuration as there are no heat exchanger losses. Most pool water contains additives like chlorine, which is highly corrosive. This corrosive property of pool water makes copper absorbers or copper pipes non-compatible. This is one of the reasons the collectors are made of plastic, which is not affected by chlorine. The same holds true for the piping, which is typically PVC.

Plastic pool heating collectors are typically mounted flat on a roof. The collectors are held in place with a set of straps that go over the collectors but are not actually attached to them. The straps are often plastic-coated stainless steel and are threaded through special clips that are bolted to the roof. This method of holding down the collectors allows them to expand and contract on the roof without binding.

Air Collectors

Up to this point, all the kinds of collectors we have talked about have used a liquid as the heat transfer mechanism. Air can also be used as the heat transfer mechanism in a solar collector. Air collectors are flat plate collectors and share all the same characteristics of liquid type flat plate collectors in size and construction. Instead of an absorber plate made of copper piping and copper fins, the absorber plate in an air collector is typically made of a solid sheet of aluminum. The aluminum absorber plate is coated with a selective surface or black paint and is usually dimpled to increase efficiency. When the sun shines on the absorber plate, it gets hot. Air is drawn from the building and is blown across the back of the absorber plate and heated. The hot air is then delivered to the building through ductwork. A blower circulates the air through the system.

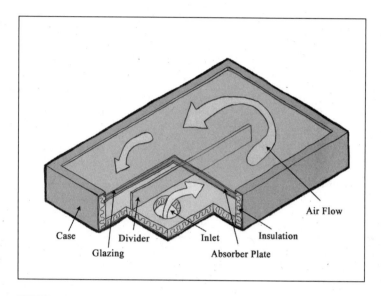

FIGURE 3.11
Air collector

Another kind of air collector is an aspirated collector. This collector is an unglazed flat plate collector. It draws fresh air through tiny holes that are punched in the absorber plate. This fresh air is heated as it passes through the collector and is delivered into the building. This kind of collector is used for heating make-up air in buildings that require many air changes per day.

Chapter 4

OTHER SYSTEM COMPONENTS

THIS CHAPTER WILL DETAIL a number of the other components that are used in solar water heating systems. This is not an exhaustive list of components, only those that are most commonly used. Components that are designed for solar are typically preferable.

Storage Tanks

Because of the path of the sun, there is only a limited time in which we can harness solar energy. In addition, we aren't willing to use hot water only when it is sunny. For instance, we mostly shower in the morning or at night. Therefore, all solar water-heating systems will require some form of storage tank, where the solar heated water is stored until needed. These tanks range in size from 40 gallons capacity to 120 gallons. Determining the size of the storage tank will be covered in Chapter 7. As with most things, there are a number of options to consider when choosing a storage tank. The type of system you are planning will determine some of the specifications of your storage tank.

But first, a quick discussion on tanks in general. The most common and traditional tank used today is a steel tank. The steel tank is encased in foam insulation to reduce heat loss and has a light-gauge steel jacket on the outside to protect the insulation. It looks just like a regular water heater. The inside of the tank should be coated with an enamel layer, often called glass lining, that is typically baked on. This lining helps reduce corrosion and significantly prolongs the life of the tank. High quality steel tanks should also be fitted with an anode, or sacrificial, rod which is screwed into a fitting on the top of the tank and extends down into the tank. An anode rod helps reduce tank corrosion by rusting before any of the system components do. Anode rods actually wear away, and their life expectancy varies depending on the conditions at your location. The anode rod should be checked every five to ten years. Steel tanks typically last 15 to 30 years, depending on the environment at your location and the quality of your water.

Fiberglass tanks are the new kids on the block. Modern developments in fiberglass technology have enabled engineers to create a cost-competitive alternative to steel tanks. These tanks are constructed much like the steel tanks mentioned above except the tank itself is made of fiberglass and the jacket is made of plastic. They have a huge advantage over steel tanks because they will not deteriorate because of rust or corrosion. When using fiberglass tanks, a few special precautions should be made. First, when screwing fittings into the tank, it is important not to overtighten the fittings. Many installers tend to crank fittings very tight when working on steel tanks, and there is no problem with that, but on fiberglass tanks you can actually break the fitting if too much force is applied to it. So be sure to use plenty of dope, sealant, or Teflon tape on the fitting and do not overtighten. Second, a vacuum breaker should be installed on the top of the tank to facilitate safe drainage. Whenever a tank is being drained and the tank is sealed at the top (all faucets above it are closed) a vacuum is created at the top of the tank. This is no problem with a steel tank, as steel is very strong and rigid. While a fiberglass tank is very strong in relation to outward pressure, it is weak when it comes to the inward pressure that would be caused by having a vacuum inside the tank.

Please be aware that at the time of this writing, no plastic storage tanks have proven to stand the test of time for solar water heating. Most types of plastic will deteriorate under the constant hot conditions. Do not use plastic or rubber lined tanks for any solar thermal application. However, tanks with a plastic exterior and a fiberglass interior are acceptable.

TANKS WITH BUILT-IN HEAT EXCHANGERS

Tanks with built-in heat exchangers are special storage tanks that are used primarily in solar water heating systems, making them a very specialized product. There are two typical configurations. The first has an internal heat exchanger that is removable and is accessible through a bulkhead fitting on the side of the tank. The other is to wrap an exchanger around the outside of the tank, sandwiched between the tank and the tank insulation. The big advantages of these tanks are: 1) installation is simplified because you don't have to do the extra work of mounting and plumbing an external heat exchanger, and 2) the system will only require a single main circulating pump. They can also save space where quarters are tight.

There are also some disadvantages. First, this method of heat exchange is usually less efficient than using an external heat exchanger. The

FIGURE 4.1

Solar storage tank with internal heat exchanger

FIGURE 4.2

Solar storage tank with wrap-around heat exchanger

Drawings by: Lief Ramlow

wrap-around method is more reliable but the efficiency is lower. The in-tank exchangers have higher initial efficiency, but they tend to get covered with sediment and become less efficient over time, especially in hard water situations. A major disadvantage of both kinds of internal heat exchanger-tank combinations is that when the tank fails, you have to replace both the heat exchanger and the tank because they are one unit.

FIGURE 4.3

Drainback tank with internal heat exchanger

These tanks would be made exclusively of steel.

TANKS WITHOUT BUILT-IN HEAT EXCHANGERS

Storage tanks without built-in heat exchangers are by far the most common kind found in solar water heating systems. These can be made of steel or fiberglass. To function as a solar hot water storage tank, the tank has to be insulated and have suitable fittings to allow proper connections of the water supply, the heat exchanger, and a drain. A number of manufacturers build special storage tanks designed for this type of application, similar to their water heating tanks but with extra ports added to make it easy to install the solar plumbing. It is also possible to use a traditional water heater tank. I typically use an electric water heater tank as my solar storage tank. These tanks are readily available at cost-competitive prices and come in both fiberglass and steel. I do not recommend using a gas water heater tank because most have a flue down the center, which can lead to more standby heat loss, as well as providing more surface area to develop leaks. Sometimes folks want to use a used tank when installing a system. If you choose to do that, you should inspect the tank very carefully for corrosion. In most cases the used tank was taken out of service either because it leaked or was old and ready to leak. I do not suggest using used storage tanks.

DRAINBACK TANKS

Drainback systems need specialized reservoir tanks in addition to the storage tanks. They are like a small solar storage tank and are usually around 20–40 gallons. Some drainback tanks have a heat exchanger built into them. Many have a sight glass built in to the side of the tank for monitoring the liquid level inside the tank. These specialized tanks are available from collector manufacturers and solar specialty shops. Like any other tank, you want to look for glass-lined steel or fiberglass tanks with adequate insulation.

SOLAR SPACE HEATING STORAGE TANKS

Solar space heating systems are just overgrown solar water heaters. Being overgrown, they need an overgrown storage tank — think big. The size will vary depending on house size and climate. It is not uncommon for these storage tanks to be in the 300-gallon to 500-gallon range. To achieve the required storage size you can either use one large tank or several smaller tanks plumbed together. One large tank will always be preferable. These tanks are made of either steel or fiberglass. There have been many attempts over the years to build storage tanks in place and line them with plastic or rubber. These do not work and will fail. The liner in these tanks is not designed to withstand temperatures commonly encountered in the system. They will often become brittle and eventually leak and crack.

One big problem with large storage tanks is getting them into a building and through doorways. Some manufacturers have made tanks that come in pieces and are assembled in place. These are usually made of fiberglass and either bolt together or are cold welded with epoxy. I use the cold-welded kinds that are made of heat-resistant fiberglass.

Heat Exchangers

Liquid-to-liquid heat exchangers are used in a solar water heating system to transfer the heat from the solar fluid

FIGURE 4.4

Plate heat exchanger

into the domestic water. Some heat exchangers are single-walled. A single-walled heat exchanger has a single membrane between the two fluids. This membrane could be made of copper, stainless steel, or (in very specific circumstances) Pex. Single-walled heat exchangers are the most efficient. Double-walled heat exchangers are also used in solar water heating systems, and some municipalities require them whenever domestic water is involved in the exchange. Double-walled exchangers afford an extra layer of protection to the potable water if the heat exchanger ever developed a leak. These rules requiring double-walled heat exchangers are a carry-over from the early years of our industry when toxic liquids were used as heat transfer fluids. There is no need for a double-walled exchanger if a non-toxic solar fluid is used.

Heat exchangers can be constructed using either pipes or plates. Two fluids are passed next to each other, separated by a membrane. Heat transfers across the membrane from one fluid to the other. The two fluids always flow in opposite directions

FIGURE 4.5

Tube-in-shell heat exchanger

when passing through the heat exchanger, as this is the most efficient way to transfer heat. A plate heat exchanger consists of a number of plates spaced apart and capped around the sides. Separate waterways are designed into them to allow the different fluids to pass in adjoining spaces. This type of exchanger can pack a large amount of surface area into a small package, and heat transfer is directly related to the amount of surface areas that are available. It is hard to have too much heat transfer surface area. Plate heat exchangers always require a pump on both waterways, as they will not thermosiphon.

A tube-in-shell heat exchanger is essentially a smaller pipe (or pipes) inside a larger pipe or tube. One fluid is circulated through the inner pipe and the other fluid circulates through the outer pipe. The material of the inner tube separates the fluids. Tube-in-shell heat exchangers come in either single wall or double wall versions. These heat exchangers can be configured in straight lengths or coiled. The coiled configuration always requires a pump like the plate exchangers, but the straight ones will thermosiphon on the waterside.

Pumps

PUMP CONSTRUCTION

It is important that all the components within a solar loop are made of compatible materials. When different materials are mixed in a plumbing circuit, galvanic reactions can take place that speed up oxidation and deterioration of the components. I suggest using brass, stainless, or bronze pumps in the solar water heating system, both in the solar loop and for all potable water plumbing. Never use a cast iron pump in any of these applications. The impellers within the pump are always made of stainless steel or plastic, so they are not a problem. It is the pump housing that is critical.

AC PUMPS

The traditional type of pump used in a solar water heating system is a 120 volt AC pump. These pumps are readily available in a variety of sizes and are also used in traditional hydronic heating systems. Because these pumps are mass-produced, they are less expensive than specialty pumps designed specifically for the solar industry. There are many manufacturers of these pumps and they come in a variety of configurations. These pumps are very reliable

and last for a long time. Because of their low price, many of these pumps are not field-repairable, so when they break, they are just replaced. A few manufacturers make pumps called cartridge circulators. These have a cartridge inside the pump that can be replaced without having to take the whole pump housing (called the volute) out of the system when doing a repair.

DC PUMPS

DC pumps come in two configurations: brush types and brushless types. This designation relates to the type of

FIGURE 4.6
High head AC pump

motor that drives the pump. Brush-type motors use brushes made of carbon that contact the commutator, which is a cylinder in the motor. These brushes wear away over time and need to be replaced periodically. Brushless motors are electronically commutated, and obviously have no brushes. Brushless motors work better in photovoltaic (PV) direct applications, but brush-type motors can work fine. Brush-type motors have a harder time starting up when powered directly from a PV panel. They will start, but not as quickly as brushless ones. A lin-

ear current booster (LCB) can be installed between the PV collector and the brush-type motor to help the brush-type motor start more easily. LCBs need to be properly sized to match the PV collector and the pump, so do your homework if using an LCB.

There is another type of brushless pump that doesn't actually have a conventional motor. This pump uses electronics to spin the impeller. These are small circulator pumps that are used in solar water heating systems where low to moderate flows are required. Some models of these 12 volt DC pumps can run directly off a PV collector while others run off a 12 volt DC power supply. They are great little pumps that have a good track record.

FIGURE 4.7

DC Pump

Piping and Pipe Insulation

PIPING

The piping in a solar loop is subjected to a wide range of temperatures, varying between 300°F. to −30°F (or lower). Copper tubing is the best kind of pipe to use for the solar loop. It can withstand this temperature range and is very durable and easy to install. Copper tubing comes in various grades and is classified by wall thickness and rigidity. Soft copper can be bent while hard copper is very rigid. The heavier the wall thickness, the more rigid it is. Type M copper tube is thin-walled, type L is medium-walled, and type K is heavy-walled. I suggest using type L for the solar loop piping. Use type K for underground piping runs. I suggest using hard copper in all instances except underground piping runs. All types of copper pipe have the same exterior dimensions, so all fittings are made the same and come in one standard size for use with all types.

It is important to use only copper pipe for the hot supply pipe in the solar loop. People often want to use a flexible product like Pex tubing or rubber hose. These products will not last very long and will deteriorate well before the system wears out. I have replaced a lot of hoses that were used in the 1980s. Not a single installation from those days that used rubber hose is still operating (to my knowledge) in Wisconsin. Pex tubing will fail in a very short time. I have seen several attempts to use Pex tubing for the hot supply line and the failures were always within the first year. It is possible to use Pex tubing for the return line back to the collectors, but it is important to terminate the Pex at least

10 feet from the collectors. The Pex must never be used for an outside pipe run unless it is buried. I suggest that you only consider Pex for use in the solar loop when burying the return line out to a ground-mounted array.

PIPE INSULATION

Pipe insulation comes in a wide variety of materials and specifications. Most kinds will not withstand the temperatures experienced in a solar loop; most plastic or rubber pipe insulation will melt right off. Only a few kinds of pipe insulation will work, and these are made specifically for high temperature situations. Fiberglass pipe insulation is great for all interior pipe runs, but is unacceptable for exterior runs or buried runs because it will soak up moisture and lose its insulation value. Products like Rubatex and Armaflex will work, but these kinds of insulation will eventually become hard and brittle and can wear away, especially when exposed to the elements. HT Armaflex will remain flexible for a long time and is an acceptable brand to use on solar loops for both interior and exterior applications. It is always best to place a jacketing material on exterior insulated pipes. It is a tough world out there and exterior insulation is subjected to UV radiation, rain, freezing conditions, insects, and birds. Installing an insulation jacket on all exterior pipe runs will greatly prolong the life of the insulation. Both PVC jacketing and aluminum jacketing are commonly used.

Solar Fluids

The best solar fluid is plain water. All other fluids are less efficient at transferring and holding heat. Because water is the best heat transfer fluid, many solar water heating system designs have attempted to use it as the heat transfer fluid. The good and bad traits of these different types of systems will be discussed in Chapter 6. Those discussions clearly point out the advantages and disadvantages of these various approaches and designs.

Closed-loop antifreeze systems use some type of fluid that will withstand freezing conditions. Many different fluids have been tried. Some examples are: propylene glycol-water mix, ethylene glycol-water mix, synthetic oil, and silicone oil. Of all the examples, only the propylene glycol-water mix is acceptable. The ethylene glycol-water mix is toxic; it will also deteriorate very quickly and damage your system. The synthetic and silicone oils require specialized components within the solar loop because the oil will dissolve all rubber and plastic components and seals, and they are just poor conductors of heat.

Propylene glycol-water mixes are the industry standard for heat transfer fluids in solar water heaters. Propylene glycol is essentially non-toxic and is a relatively stable product, even at high temperatures. Be aware, though, that not all propylene glycols are created equal. These glycols come in various formulas and you need to carefully choose the appropriate formulation. You want to choose a product that is formulated to withstand at least 300° F without breaking down. Most commercially available high-temp glycol products also have additives mixed with the glycol to help stabilize the product and make it more compatible with system components.

You never use 100 percent glycol as your solar fluid: always mix it with a certain proportion of water. You want to have the highest percentage of water in the mix as possible and still have the protection you need for your particular climate. Dilution charts provided with the product show the necessary concentration of glycol in the mix to protect the system at various temperatures. Figure 4.8 is an example of a dilution chart.

These dilution charts show two types of protection: freeze and burst. As glycol-water mixes cool, the fluid becomes thicker and harder to pump. At a certain point the fluid will have a jelly-like consistency but will continue to protect the piping from bursting. The freeze temperature on the chart shows the temperature at which the fluid turns into jelly. The burst temperature on the chart is the temperature at which the fluid will begin to freeze solid and burst a rigid pipe. For instance, a high quality glycol like DowFrost HD at a 50/50 mix of water and glycol will provide freeze protection down to -34°F and burst protection down to -70°F. When choosing a dilution for your climate, you should choose one that is at least 5°F below the lowest temperature you might experience at your location. Note that when the temperature gets near the freezing point of your dilution, the solar water heating system will stop working, but you will still be protected from bursting pipes to a much lower temperature. The leanest mix of propylene glycol you should consider in the mix is 20 percent glycol. The richest mix of propylene

glycol you should consider in the mix is 60 percent glycol. Remember to always use the leanest percentage of glycol that you can get away with and still have the protection you need for your climate.

It is important to note that glycol is slippery and harder to pump than water. Plus, as the glycol gets hotter it gets even more slippery and harder to pump — all the more reason to use the leanest glycol mix that will provide protection. Also note that any pump curve you look at relates to pumping plain water. Because glycol mixes are harder to pump than pure water, it is important

FIGURE 4.8

Percent Glycol Concentration Required

Temperature		Percent Glycol Concentration Required	
C	F	For Freeze Protection Volume%	For Burst Protection Volume%
-7	20	18	12
-12	10	29	20
-18	0	36	24
-23	-10	42	28
-29	-20	46	30
-34	-30	50	33
-40	-40	54	35
-46	-50	57	36
-51	-60	60	37

to consider the viscosity of glycol when sizing your pump.

Propylene glycol fluid in a solar water heater will deteriorate over time and will eventually wear out. The solar fluid has to be checked periodically. Details are given for this procedure in the maintenance section of Chapter 9. On average, a high quality and high temperature glycol solar fluid will last between 15 and 20 years. If the fluid is subjected to abnormally high temperatures or long periods of stagnation (no flow), its life will be reduced. If either of these situations occur, the

FIGURE 4.9

Expansion tank

fluid should be checked more often. When glycols deteriorate, they become more acidic, which is harmful to the system, and their freeze protection capacity is diminished.

Other System Components

EXPANSION TANKS

Liquids expand when they are heated; this expansion can be substantial. Unlike gases, liquids cannot be compressed. In a solar water heating system, the solar fluid can experience dramatic temperature changes over the course of a single day. The expansion and contraction of the solar fluid over this range of temperatures must be compensated for, or else the system would burst. An expansion tank is the component that does that compensating.

Expansion tanks contain a volume of air contained in a chamber which is open to the main circuit. As the fluid is heated and expands, pressure builds up in the system. As the pressure increases, the air in the expansion tank compresses and makes room for the expanded liquid. Commercially available expansion tanks have a bladder separating the liquid from the air. The air pressure can be adjusted in the air chamber via a Schrader valve on the bottom of the tank. The air pressure should be set to three pounds per square inch below the pressure of the system at 60°F fluid temperature. This can be measured with a regular tire gauge. See the section on system designs in Chapter 6 for pressure suggestions.

Expansion chambers that do not have a bladder should be fitted with a ball valve which should be closed during the charging and pressurizing procedure and then opened after the pressurizing is complete. These air chambers can be fitted with a Schrader valve fitted to the top of the chamber.

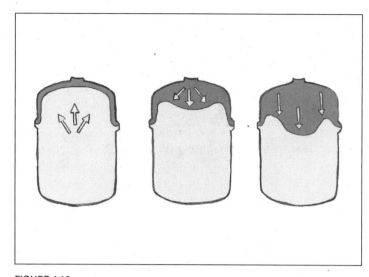

FIGURE 4.10
Expansion tank process

BULKHEAD FITTINGS

The only time you will have to make a choice regarding bulkhead fittings is when you are modifying a large tank in a space heating or large commercial system application. Bulkhead fittings are fittings that add a port to a tank, usually below water level. You can think of a typical bulkhead fitting as a large hollow bolt. The hollow inside this bolt is threaded so a pipe can be threaded into it. These fittings fit through a hole in the tank. There is usually a flange on both the inside and the outside of the fitting and rubber washers are placed between the washer and the tank. When the nut and bolt are tightened together, the washers tighten to the tank on the inside and outside, creating a watertight connection. I prefer not to use bulkhead fittings if possible and instead run the piping up through the top. It is difficult to get a bulkhead fitting to seal on a round tank. Also, avoid plastic bulkhead fittings; always use metal if you have to have one.

THERMOMETERS

It is interesting to have some thermometers installed at various locations on the solar energy system. They help you monitor the performance of the system. They are not necessary for the operation of the system, however, and I consider them a bell and whistle type of option. There are two types of thermometers to consider: mechanical or electronic. Mechanical thermometers typically mount in a well or chamber placed within the pipe that carries the liquid you want to monitor. Note that you have to plan ahead and locate the wells in the proper place as the circuit is being plumbed. Well thermometers are accurate and relatively inexpensive. Electronic thermometers use sensors

that are fixed to the pipe where you want to monitor the temperature. A well is not required, so no prep work needs to be done. An advantage of electronic thermometers is that you can often locate the temperature display some distance from the site of the sensor. People often locate the display in an area that they frequent, so it is easy to see what is going on without having to be present at the site of the sensor. Some equipment allows you to read the temperatures at several locations on one display. There are even more sophisticated temperature monitors that will interface with your computer, so you can do accurate data logging.

The most common place for thermometers is on either side of the heat exchanger on the solar loop. This placement allows you to see the temperature of the incoming solar fluid, and the temperature drop across the heat exchanger. I also like to put a thermometer on the solar storage tank, near the top, to monitor the temperature in there.

PRESSURE GAUGE

All closed-loop systems require a pressure gauge. The gauge can be placed anywhere on the solar loop, but it

FIGURE 4.11

Pressure gauge

should be in close proximity to the charging ports (see "Charging a Closed-Loop System" in Chapter 9). Pressure gauges are inexpensive, so get a good one. You can get a combination temperature/pressure gauge. These work fine, but again, get a good one as the cheap ones fall apart.

CHECK VALVES

Check valves allow the fluid passing through them to flow in one direction only. An arrow on the valve body indicates the direction of flow. Two kinds of check valves are commonly used in solar water heating systems: spring check valves and swing check valves. Both kinds have their advantages and disadvantages. I have used swing check valves exclusively for over 15 years with no failures.

Swing check valves control the direction of flow by using a swinging door inside the valve. When the fluid is flowing in the direction of the arrow, the moving fluid pushes the door open and flows through the doorway. When fluid flows in the other direction, the fluid pushes the door shut, stopping the flow. These valves are typically made of bronze, so the metal is fairly hard and very durable. When the door closes, it rests against a machined

FIGURE 4.12
Swing check valve

FIGURE 4.13
Swing check valve process

FIGURE 4.14

Swing check valve

FIGURE 4.15

Swing check valve process

surface, which is very flat and even. There are no gaskets inside the valve; the seal is simply metal against metal. These valves can be located on horizontal pipe runs and vertical pipe runs where the fluid flow is upwards. This is important! Swing check valves will not work if the flow direction arrow is pointing anywhere below horizontal (pointing down at any angle).

Spring check valves use a spring that holds a door or a ball closed. Fluid flowing through the valve in the direction of the arrow will compress the spring and allow the door to open. These valves can be positioned in any direction. They do not rely on gravity to close the door when flow stops. Most spring check valves are produced to operate in systems that use powerful pumps or high flow conditions, so the springs are often quite stiff. This can cause a serious restriction when small pumps are used, as is often the case in solar energy systems. There are some specialty spring check valves that use modified soft springs that open more easily. Some experienced installers modify the springs in commercial spring check valves by opening up the valves and cutting the springs down so they open more easily. I don't use them.

Some installers have used a motorized valve to close the system down. While this can work, I do not suggest it because it adds an unnecessary level of complication and expense. Keep it simple!

DIFFERENTIAL TEMPERATURE CONTROLLER

Differential temperature controls are electronic devices that can compare the temperatures of two remote locations. They have a built-in microprocessor that can turn on a pump when the temperature is warmer in one location than it is in the other. These controllers use electronic sensors to measure the temperatures at each location. They include a relay switch that turns the pump on or off. Most controllers are adjustable, so you can set the temperature differential that the logic will use to operate the pump. Some controllers also have a high limit feature that turns off the pump when the storage temperature reaches a pre-set temperature.

SNAP DISC THERMOSTATS

A snap disc thermostat is an inexpensive switch that turns a circuit on at a preset temperature. Most snap disc thermostats are not adjustable. These thermostats are not very accurate or

FIGURE 4.16

Differential temperature controller

FIGURE 4.17

Snap disc thermostat

FIGURE 4.18

Aquastat

FIGURE 4.19

Automatic air vent

durable and should not be used in a solar water heating system.

AQUASTAT

An aquastat controller is a switch that is usually adjustable and turns on and off at different temperatures. It is a more accurate and more expensive type of thermostat switch that measures temperature and can adjust it. An aquastat is better than a snap disc switch, because a snap disc can be set for one temperature only. With the aquastat you can set the temperature at which it will go on and off, and also set the differential. Part of the aquastat is a sensor and the sensor is actually like a thermal couple. A thermal couple is a temperature sensor that creates its own millivoltage; the voltage it creates is determined by the temperature of the sensor or the bulb. There are two types. One has the sensor built in to a strap that fastens on to the pipe, keeping the sensor in contact with the pipe. The other, called a remote bulb thermostat, has a small copper tube which connects the sensor to the aquastat.

AUTOMATIC AIR VENT

Automatic air vents are valves that allow air to escape from a circuit that is full of fluid. They are placed at the highest point in the system or are installed on the top of an air eliminator. These valves are especially helpful in traditional hydronic heating systems, where fresh water is continually added to the system. They can be an important component in a solar water heating system when placed on the hot water side of the heat exchanger, where fresh water is always present. Note that fresh water always contains a small percentage of dissolved oxygen, and that oxygen starts to form bubbles as the water is heated in the heat exchanger. If the bubbles are allowed to add up to a big bubble, this can impede the flow of water through the heat exchanger, especially if the heat exchanger is set up to thermosiphon on the water side.

Automatic air vents come in a wide range of designs and qualities. The cast brass units are best, and the larger units are better as well.

UNIONS

Unions are fittings that join two pipes together in a plumbing circuit. They allow the pipes to be taken apart without having to be cut. They are used on all the nipples of the collectors and are used on either side of critical parts in the solar loop, like the circulating pumps and the heat exchanger/tank fittings. The best unions are made of solid brass.

There are also specialty unions called dielectric unions. These have a rubber and plastic gasket and washer that electrically isolates the two sides of the fitting. These fittings are usually placed where connections are made to the storage tank, preventing galvanic reactions between the copper piping and the steel tank. It is a good idea to use them.

DOLE VALVE

A Dole valve is used to protect flooded systems from mild freezing conditions.

FIGURE 4.20
Dielectric union

FIGURE 4.21
Dole valve

FIGURE 4.22
Drain valve

Two Dole valves are used on each collector array, one on the bottom manifold end if flat plate collectors are used. They open a drain as the temperature drops to 44°F and this allows fresh warm water to flow through the array, protecting the collectors. These valves are used only when freezing conditions are a very rare occurrence. I do not recommend using them for a number of reasons. First, I never recommend flooded systems except ICS systems. Second, if you are in a climate that has any possibility of experiencing freezing conditions, you should choose either a closed-loop or a drainback system. However, they are often installed on ICS systems simply as a precaution.

DRAIN VALVE

Drain valves do exactly what their name implies. They allow you to drain fluid from the system. Most will be of the same quality. They are used in every type of solar water heating system and are a very common plumbing implement.

BALL VALVES AND GATE VALVES

Ball valves and gate valves are used to stop the flow of fluid in a piping circuit. The traditional gate valve has a spin handle that operates a gate that slides up and down within the valve.

These valves always create resistance within the circuit and are only intended to be operated in the fully open or fully closed position. They should never be used in a solar energy system. Ball valves have a lever attached to a ball that rotates within the valve. The ball has a hole running through the middle of it. When the ball is rotated to the open position, fluid can pass through the hole in the ball and flow can proceed. When the ball is rotated to the closed position, the hole is perpendicular to the flow and flow is stopped. Ball valves can be set in any position to regulate the flow through the circuit. Always use "full-flow" or "full-port" ball valves.

RELIEF VALVES

Pressure relief valves protect circuits from excessive pressure build-up. They have a gasket attached to the end of a plunger that has a heavy spring holding the valve closed. When pressure in the system exceeds the rating of the valve the spring compresses, the valve opens, and pressure is released. For all pressure relief valves, it is important to attach a drain pipe to the valve outlet and to run the pipe so that it terminates in an out of the way location, so if the valve opens, hot pressurized

FIGURE 4.23
Gate valve

FIGURE 4.24
Two port ball valve

FIGURE 4.25
Pressure relief valve

FIGURE 4.26
Temperature and pressure relief valve

water will not spray on anybody. You can purchase pressure relief valves that are factory set to open at a specific pressure; other pressure relief valves can be field-adjustable. This is a required valve for every closed-loop solar water heating system.

Another similar valve is a temperature and pressure relief valve. These valves open either when the pressure in the system exceeds the preset pressure, or when the temperature in the system exceeds its preset temperature. These valves are designed to be installed on conventional gas or electric water heaters to protect them from overheating or over-pressurization. These valves should never be used on the solar loop!

TEMPERING VALVE

Sometimes called mixing valves, tempering valves mix hot and cold water together so the finished product is at the desired temperature. These valves are always used with a solar water heating system to protect the hot water users from being scalded. There are three ports on a tempering valve: hot, cold, and mix. There is also a handle, which is turned to set the desired temperature of the mix. If the flow of hot liquid through the valve is below the set temperature, the fluid flows straight through the valve and no mixing takes place. If the fluid entering the valve is hotter than the set temperature, some cold fluid is added to the hot fluid to cool it down to the desired temperature. When the water in the solar storage tank is too hot to be safe, the tempering valve is used to lower that temperature. It is important to always install a check valve on the cold line going into the tempering valve to prevent heat from bleeding into the cold water line. I suggest using high quality tempering valves, as the poorer quality ones tend to be less reliable.

MOTORIZED VALVES

Often called zone valves, motorized valves turn a circuit on or off by using a motor to operate the valve. They are typically used in hydronic heating systems to control the flow of heat to various zones within a building. Some motorized valves operate on 120 volt AC while others operate on 24 volt AC, the typical operating voltage for heating controls. Note that there is always a direction of flow indicated on the valve housing. These valves should never be used on the primary solar loop.

FIGURE 4.27
Tempering valve

FIGURE 4.28
Zone valve

FIGURE 4.29

Vacuum breaker

VACUUM BREAKER

Vacuum breaker valves are used to let air into a system when the system is depressurized. They contain a plunger with a gasket on the end. When the system is under pressure, the plunger is forced against a seat, sealing the system. When the pressure is lost, the plunger drops away from the seat and air is allowed into the system through the valve. Typically found in solar swimming pool systems and some draindown and drainback systems, these valves facilitate fast system draining by allowing air to enter the system. They can also be used on the top of a fiberglass water heating or storage tank, where they help prevent the tank from collapsing when being drained.

Chapter 5

SOLAR WATER HEATING SYSTEMS

Introduction

BEFORE I START detailing the various system designs for solar water heaters, it is best to clarify some terms that are used to classify and define the systems. I will also go over some basic principles that are essential for understanding the system designs.

Be sure to read all of this chapter to fully understand solar water heating systems. Each section explains a particular type of solar energy system, telling how it works and the basic components it uses. Another reason to read each section is because certain concepts and definitions are given in early sections that are also used in following ones.

BASIC TERMS

Direct Systems

In a direct type of solar water heating system, domestic water (the water we use in our home) actually enters the solar collector, where it is heated.

Indirect Systems

In an indirect system, a heat transfer fluid (also called solar fluid) is heated in the collectors and then circulated to a liquid-to-liquid heat exchanger where the solar heat is transferred from the solar fluid to the domestic water.

Passive Systems

Passive solar water heaters circulate either domestic water or a heat-transfer fluid through the system without the use of pumps. Typically, passive systems have few (if any) moving parts and require no external energy to operate. Being very simple, they tend to be very reliable and easy to maintain. Because of their simplicity, they also tend to be the least expensive system to choose from. But because of their simplicity, they may be vulnerable to problems that other systems can overcome.

Active Systems

All active solar water heaters utilize pumps to circulate fluids throughout the system. Some form of external power is required for operation.

Open-Loop Systems

All open-loop solar water heating systems are direct systems because the water travels through the solar collector. All open-loop systems are active because they use pumps in their operation.

Closed-Loop Systems

In closed-loop systems, the solar fluid remains within a single circuit at all times. All closed-loop solar water heating systems are classified as indirect systems because the sun heats some type of solar fluid and that heat is transferred to the domestic water through a liquid-to-liquid heat exchanger. Closed-loop systems can use flat-plate, evacuated tube, or concentrating collectors. Flat plate collectors are by far the most common.

Two-Tank Systems

All systems, except batch systems, need some type of storage tank to store the water that was heated by the sun. Most systems use a storage tank for the solar heated water, and also a back-up heater which can be either a tank-type heater or an on-demand (tankless) water heater. These systems are called two-tank systems. Some systems attempt to use the solar storage tank as the back-up heating system as well by including an electric heating element or a gas burner to heat the water when there is insufficient solar energy available to bring the water up to the desired temperature. These systems are called one-tank systems.

The problem with one-tank systems is that if hot water is used during the evenings and depletes the storage tank of hot water, the electric or gas heat source will heat up the tank. Then when the sun comes out the next day, the tank is already warm and the efficiency of the solar water heater is greatly diminished. I typically install two-tank systems because you usually need a second tank to make suitable storage for the collectors. All of the illustrations of systems are of two-tank systems.

BASIC PRINCIPLES

Hot Water Rises

A principle of physics we see in action in solar water heaters is that of heat rising. When a fluid is heated, its molecules expand. As molecules expand, they become lighter and rise above any surrounding, cooler (denser) molecules. If we observe a tank of water with some water in it that is hotter than the rest, the hottest water will rise to the top of the tank. This phenomenon is called stratification. This principle will be used to our advantage in most solar water heating systems.

Water Expands When it Freezes

Compounds contract as they cool. As they contract, they become heavier. Eventually, a liquid can become so dense that it becomes a solid. This is the opposite of what happens as compounds are heated. Water is unique among most compounds. Most compounds are densest in their solid state.

Water is densest just before it freezes (becomes a solid) and then expands again as it gets colder. If water acted like most other compounds, it would sink instead of floating when surrounded by liquid water. Life as we know it would be vastly different, if it would be here at all, if this happened. Just think — if ice sank instead of floating, as lakes froze, the ice would sink and crush all the fish. All our lakes, rivers, and oceans would freeze solid and probably never thaw out. This quirk of physics, where ice floats on water, allows life as we know it to exist.

This expansion is a very powerful force. Freezing water has shaped the earth by breaking apart rocks into soil. Water gets into cracks in rocks, and when it freezes and expands can literally explode a rock apart. This can be a bad thing in a solar energy system. If water in a pipe freezes, it expands. Most of the time, this expansion exceeds the limits of what a pipe can take, and the pipe breaks.

Water Quality

In most locations, our water supply is considered "hard." Hard water is water that has dissolved minerals in it. Most of us are familiar with water softeners: they remove these minerals.

When hard water is heated under certain conditions, these dissolved minerals can precipitate out of the water and solidify. People who have hard water often experience problems with their conventional water heaters. Electric water heaters tend to go through an excessive number of heating elements. Gas water heaters tend to build up mineral deposits on the bottom of the tank (where it gets the hottest) and this sometimes causes the heater to make various noises when the burner is ignited. Having hard water is also one of the main causes of premature tank failure. This same thing can happen in a solar collector. Mineral build-up inside a collector will decrease collector efficiency by depositing a layer of minerals on the inside of the collector. This mineral build-up acts like a layer of insulation on the inner surface of the waterway and slows the transfer of solar heat to the water. In collectors with small waterways, this mineral deposit can actually plug the pipes and render the collector useless.

RECOMMENDATIONS

While a number of systems have been used in the past, there are really only three domestic water heating systems that I would ever recommend: ICS, drainback, and closed-loop antifreeze. These three are the only systems that have proven their reliability over the past 20 years. Some of the systems that are described in the following pages are indentified solely for the purposes of education and repair. If you are a solar professional, you need to be prepared if you see one of these systems and are called on a service call.

Integral Collector Storage Systems

Integral collector storage (ICS) solar water heating systems are considered passive because they require no pumps of any kind for operation. They are considered direct systems because the domestic water actually enters the collector. The ICS unit is typically plumbed in series between the cold water supply and the conventional water heater. Whenever a hot water tap in the dwelling is opened, cold water from the supply enters the ICS collector and forces the solar heated water stored there into the conventional or back-up water heater. If the water from the ICS collector is hotter than the setting on the back-up heater, that heater will not activate. If the water from the ICS collector is warmed but is below the temperature setting of the back-up heater, the back-up heater will then have to add only enough heat to bring the water up to the pre-set temperature. If no solar heating has taken place and all the water in the ICS collector is cold, the back-up heater will have to deliver the whole load.

Because water is plumbed through the ICS collector, this type of system is only suitable for climates or seasons where there is no chance of freezing conditions being encountered. ICS collectors are very popular along the extreme southern parts of the US and in tropical climates. ICS collectors have also been successfully used for summertime use only, typically in conjunction with vacation homes and recreational facilities like parks where they are only used during the summer months and are drained the rest of the year.

ICS Collector

Cold Water Supply

Hot Water Load

Pressure & Temp Relief Valve

Boiler Drain

Ball Valve

Tempering Valve

Back-up Water Heater

FIGURE 5.1

ICS system

Because of their simplicity, ICS collectors are among the least expensive solar water heating systems available. There are no pumps, extra storage tanks, controllers, or other components needed. A complete system consists of the ICS collector and piping. Because of its simplicity and low initial cost, people who live in climates where freezing conditions do occur are sometimes tempted to use this system. They often think that if freezing conditions are forecast, they will simply drain the water out of the collector and piping and everything will be all right. In theory, this assumption is correct. In reality, Murphy's Law — if anything can go wrong, it will — often comes into play. Actually, all ICS collectors can withstand mild freezing conditions. The mass of the water in the collector will keep it from freezing for quite some time (depending on the temperature). It is the piping going to and from the collector that is vulnerable. A well insulated three-quarter inch copper pipe will freeze in less than 5 hours at 29°F.

Most ICS solar water heaters do not suffer because of hard water. Only in areas with extremely hard water is there a potential problem. Most ICS solar water heaters have large tanks and waterways, decreasing potential mineral build-up problems. Also, the tanks do not tend to get excessively hot, decreasing mineral deposit build-up.

If you live in a climate that does not experience freezing conditions and you do not have excessively hard water, this is the system for you. If you only need hot water during the summer months, for a campground or summer cottage, for example, the ICS solar water heater is a viable option for you as well.

Because ICS systems actually store water within the collector, these collectors weigh the most of any type of solar collector. Water weighs over 8 pounds per gallon, so a 40-gallon ICS solar water heater can easily weigh over 500 pounds. Provisions may need to be made if this type of collector will be roof mounted. Most buildings can support an ICS solar water heater with no modifications, but be sure to check out your situation before going ahead. It is no fun having a collector full of water — and perhaps a few installers — come tumbling through your roof.

Thermosiphon Systems

Thermosiphon systems all utilize the principle that hot water naturally rises. In these systems the storage tank is located above the collectors so that as fluid is heated in the collectors it rises to the storage tank. These systems are constructed similarly to ICS systems, except that you can construct them out of individual components instead of a single collector unit. Two systems utilize the thermosiphoning principle. Both are considered passive systems because they don't require pumps.

DIRECT THERMOSIPHON SYSTEMS

Direct thermosiphon systems typically use a flat plate collector and heat your domestic water directly by circulating it through the collector. They also use a storage tank to hold the solar heated water. In most of these, the bottom of the storage tank is connected to the bottom of the collector with a pipe, and the top of the storage tank is connected to the top of the collector with a pipe. Because the tank is located above the collector and is filled with water, the collector is always full of water. When the sun comes out, it heats the water in the collector. At this point, the water in the collector is warmer than the water in the storage tank. Because the water is warmer than the water in the storage tank, it will rise up to the top of the storage tank through the pipe that connects the top of the collector with the top of the storage tank. As this water rises within the system, cooler water from the bottom of the tank is drawn down to the collector through the pipe that connects the bottom of the storage tank with the bottom of the collector, setting up a circulation without the use of pumps. The most efficient way to heat this water would use flat plate collectors. However, because the system is direct, you will be sending hard water through the collector. In most cases this is a bad idea. Flat plate collectors have small riser tubes that are easily clogged with mineral deposits. If this system were to be installed, a water softener located prior to the collector would be essential. While this is one of the most simple systems, if your location permits you to use this system, you would be better off if you just installed an ICS system.

INDIRECT THERMOSIPHON SYSTEMS

Some indirect thermosiphoning systems have been manufactured over the years. Like the thermosiphon system described above, the indirect system also has the storage tank mounted above the collector. Instead of circulating the domestic water through the collector, a heat transfer fluid is used, which eliminates problems caused by hard water. When the solar fluid is warmed in the collector, it rises up naturally to the level of the storage tank where a heat exchanger transfers the heat from the solar fluid to the domestic water. A chamber where the solar fluid is circulated often surrounds the storage tank. By circulating the hot solar fluid through this chamber, the water tank it surrounds is heated. The system can also be installed using a sidearm tube-in-shell heat exchanger like that seen in Figure 5.2. The collector in this type of system is protected from freezing because an antifreeze solution is used as the solar fluid, but the piping

FIGURE 5.2

Indirect thermosiphoning system

Hot Out

Cold In

Heat Exchanger

Solar Thermal Collector

Solar Storage Tank

Back-up Water Heater

Boiler Drain
Ball Valve
Pressure & Temp. Relief Valve
Expansion Tank

Thermometer
Pressure Gauge
Check Valve
Pressure Relief Valve
Tempering Valve

serving the storage tank could still freeze. I have seen this system successfully installed: if the storage tank is located in a heated area, you eliminate freezing problems. However, configuring the system in this manner is typically not suitable for most locations. Where it has worked, the collectors were mounted on the overhang of a porch, and the storage tank was located on the third story of the home. Typically, the utility room is located in the basement and not on one of the upper levels. Most homeowners aren't willing to sacrifice the amount of area that it takes for the rest of the system. For this reason, indirect thermosiphoning systems are usually not the best option.

Open-Loop Systems

FLOODED OPEN-LOOP SYSTEMS

A flooded collector open-loop system has water in the collectors and piping at all times. This system is composed of a storage tank, a collector (or collectors), a circulating pump, a controller, and piping. The collector is plumbed directly to the storage tank. When solar energy is available for heating, the pump turns on. Water is circulated from the bottom of the storage tank to the collector, where it is heated. The water, after being heated directly in the collector, is then circulated back to the storage tank. At the end of the day, the pump simply turns off and the circulation is stopped. These systems can use a microprocessor-based controller with sensors to control the operation of the pump. A simpler system uses a photovoltaic (PV) solar collector to convert solar energy directly into electricity to operate the circulating pump. A PV-powered system eliminates the need for a controller because the only time the pump will run is when solar energy is available.

These systems typically use flat plate collectors, but can also use evacuated tube or concentrating collectors. While they are more complicated than passive systems, they are still less expensive and simpler than most other active systems. Because most of these systems use flat plate collectors, there are more options for collector placement than ICS systems.

Because these systems are flooded, they are vulnerable to both freezing conditions and mineral build-up in the collectors. They have no freeze protection and should not be used in locations that ever experience freezing conditions. In addition, hard water is a concern and a water softener located before the collector is essential.

RECIRCULATION SYSTEMS

Recirculation systems are exactly like open-loop flooded systems but they have the added feature of being able to recirculate warm water from the storage tank throughout the system if freezing conditions are experienced. Recirculation systems must be micro-processor controlled. If freezing conditions are experienced, the controller turns the pump on and water is circulated throughout the system, reducing the likelihood of freezing. Many differential temperature controllers have a recirculation function. This system really only works during mild and

FIGURE 5.3

Flooded open-loop system

occasional freezing conditions, so where extended freezing conditions are experienced, it is not appropriate. Another negative for these systems is the fact that heat stored in the storage tank is lost during the recirculation process, seriously decreasing their overall efficiency.

Draindown Systems

All draindown solar water heating systems are direct systems because the domestic water is heated directly in the collector. They are made up of a collector (or collectors), a storage tank, and piping to connect the collector and storage tank to the home piping, and may also use an electronic control, various valves, and a pump

or pumps. Typically, when solar energy is available, a specialized valve, like a Sunspool, opens and allows household water to fill the system. Then a pump, controlled by a microprocessor with sensors, turns on and circulates cold water from a storage tank to the collectors. The water is heated as it passes through the solar collector and then

Collector

From Water Softener

To Water Heater

Automatic Air Vent
Dole Valve
Boiler Drain
Ball Valve
Pump
Check Valve
Differential Temp. Controller
Pressure & Temp. Relief Valve

Sensor

Solar Storage Tank

FIGURE 5.4
Recirculation system

FIGURE 5.5
Draindown Valve

returns to the storage tank. Some systems use a photovoltaic collector to power the circulating pump, thus eliminating the need for the controller and an external energy input. When the sun sets or is blocked by clouds, the pump shuts off and the water that was in the collector(s) is allowed to drain down out of the collector and piping. This small amount of water is piped to a floor drain and is lost.

Draindown systems are very efficient because the water is heated directly, as in an ICS system. Because draindown systems use a flat plate collector, these systems are even more efficient than ICS systems. But like any system that heats the water directly, there is the problem of mineral build-up in the collectors. As mentioned above, this is a serious consideration and if there is any mineral content to your water, this type of system is not recommended. The specialized valves necessary for this type of system have been prone to failure. Because of this type of system's tendency to freeze, it is not recommended for climates where freezing conditions may occur more than once or twice a year.

Drainback Systems

Drainback systems are one of the three most popular types of solar energy systems installed worldwide. They are an excellent choice for all climates except those that experience severe or extended cold conditions, or where a significant amount of snow is expected annually. In hot climates that experience few, if any, freezing episodes per year, they are the best choice.

Drainback systems work very well in warm climates because when the storage tank gets heated to its maximum desired temperature (its high limit), the system turns off and all the fluid drains out of the system, which prevents the solar fluid from degrading due to overheating. This is especially the case when systems experience idle periods like vacations during the summer months in warm or hot climates. It is also the case in large commercial water heating systems in warm and hot climates where the heating load is not consistent. Because water — the best heat transfer fluid — is most often the solar fluid, these systems perform very well and at high efficiency when designed and installed properly.

Drainback systems are classified as indirect solar water heaters because the domestic water is heated by a solar fluid and heat exchanger and not heated in the collector. They are classified as closed-

FIGURE 5.6
Drainback system

loop because the solar fluid remains within a single circuit at all times. They are classified as active systems because they always use a pump or pumps to circulate fluids through the system.

The major limitation of drainback systems is their inability to prevent freeze-ups in climates with extended cold or snowy conditions. Here in central Wisconsin where I live and work, drainback systems are prone to failure. Some installers like to add antifreeze to the solar fluid used in drainback systems to extend their range northward. I am not a fan of this practice because it defeats one of their big advantages, the use of pure water as the solar fluid. Also, every time the system drains back, a thin film of antifreeze is left inside the collector, which dries and leaves a small residue. Eventually the film will build up enough to degrade the collector and diminish its efficiency. If you live in a cold climate, use closed-loop antifreeze systems instead of drainback systems.

The only reason to add glycol to the solar fluid is to prevent freezing caused by the unlikely event that the pump would stay running because the controller is malfunctioning. In this scenario the solar fluid would always be moving, and moving water is hard to freeze unless it is very cold outside. If it gets that cold where you live, you should be using a closed-loop antifreeze system.

Another limitation is that the collectors have to be located above the drainback tank. This eliminates the option of ground-mounted arrays. It should also be noted that the high-head pumps required to operate these systems use considerably more electricity than the small pumps used on closed-loop antifreeze systems. Also be aware that pump selection is more critical in drainback systems and there are height restrictions on how high a high-head pump can lift water. Unless the circulating pump has a motor over one-half horsepower, the distance between the water level in the drainback tank and the top of the collectors must be less than 28 feet.

Drainback systems use flat plate or evacuated tube collectors to heat a solar fluid, usually distilled water, which circulates through them. Two insulated pipes connect the collectors to a specialized tank called a drainback tank. A high-head pump is installed on the pipe that feeds the collector array. A differential temperature controller turns the circulating pump on and off. When on, solar fluid is pumped from the drainback tank to the collectors, where it is heated. The solar fluid then drops back down to the drainback tank completing the circuit. When the system turns off, the pump stops and all the solar fluid in the collectors and in the piping drains back into the drainback tank. Drainback tanks are relatively small, so they do not store much heat. When the system is operating, various methods, described below, are used to transfer the heat from the drainback tank to a solar storage tank.

Drainback systems are used for residential water heating systems and residential space heating systems. They are also used for commercial water heating systems and space heating systems. Most manufacturers offer drainback kits that are pre-engineered to take the guesswork out of designing a system. They work great for the majority of these installations. Since most solar water heating systems are the same size and installed the same way, these kits are fine for most situations. I highly recommend these kits if purchased from a reputable manufacturer.

Sizing a drainback system is exactly the same as sizing a closed-loop antifreeze system except that you can deduct five to ten percent from the array size to compensate for the water's

higher efficiency as a heat exchange fluid compared to glycol. System sizing is covered in Chapter 7, but only a handful of system sizes will satisfy the majority of domestic water heating demands, so system sizes and components are fairly easy to choose.

To make sure that drainback systems function properly, a few rules must be adhered to:

➤ The system must be installed to facilitate fast, complete drainage when it turns off.

➤ The collectors must be mounted so that they drain towards the inlet of the array; this would be the bottom manifold where the solar fluid enters the collectors.

➤ They should be mounted at a 15-degree angle sloping toward the feed inlet.

➤ Collectors should never be mounted so that the riser tubes of the absorber plate are horizontal, as they will sag over time and prevent proper drainage. If sagging occurs, water can be trapped in the pipe and burst it during freezing conditions.

➤ All piping that is not in conditioned space must be sloped at a 15-degree angle toward the drainback tank and all horizontal pipe runs must be supported at least every four feet to prevent any sagging, which would inhibit proper drainage.

➤ All horizontal pipe runs within the conditioned space can be run at a minimum ten-degree slope toward the drainback tank.

➤ Use a pair of 45-degree elbows instead of a 90-degree elbow whenever practical to facilitate faster drainage.

➤ The minimum pipe size for drainback systems is three-quarter-inch hard copper.

Drainback tanks should always be located within the conditioned space. They can be located at the highest point within the conditioned space to reduce head pressure. This is an important consideration when the system is installed in a two-story house and the storage tank is in the basement. Drainback tanks should be unvented except on large space heating systems or large commercial systems. All drainback tanks should be fitted with a sight glass to monitor fluid levels within the tank. The only exception to this would be where a low pressure-drop flow meter is used in the feed line as the sight glass. To size a drainback tank, calculate all the liquid that would fill the collectors and all the piping above the drainback tank and add four gallons. These tanks should be well insulated to prevent heat loss.

The solar loop pump must be a high-head pump of sufficient size to pump the water from the drainback tank to the top of the collector array. When calculating the head, measure from the bottom of the drainback tank to the highest point of the collector array and add four feet. Your pump must be able to exceed that head. Once all the piping is full of solar fluid, the pump does not have to work very hard because gravity pulling the fluid back down the return line helps pull fluid up the feed line. It will take a 120 volt AC pump to do this job, so the system must be powered by 120 volt AC and use a differential temperature controller to turn the pump on and off. These systems cannot be PV powered because when the system turns on, the pump must start with full force to overcome the head pressure, and PV-powered pumps don't work that way. The pump should be located at least three feet below the bottom of the drainback tank and be located in a vertical pipe, pumping up to the collectors.

Drainback systems include a storage tank to store the solar heated water for later use. There are various methods of getting the heat from the drainback tank to the storage tank, all of which use a liquid-to-liquid heat exchanger. One method circulates the hot solar fluid through an in-tank heat exchanger (including the wrap-around type). In this design only one pump circulates the hot solar fluid throughout the system. This type of system is called a single-pumped system because there is only one pump. Another method uses an external heat exchanger mounted to the storage tank. This method operates similarly to the example above and can be seen in Figure 5.6. In this case a second pump may be needed to circulate the water from the storage tank, through the heat exchanger, and back to the storage tank. This style is called a double-pumped system because it uses two pumps. Both pumps are controlled by the differential temperature controller and turn on and off at the same time. The heat exchanger pump should be very small because the flow through that circuit should be slow, and because there is very little total head to overcome. Too large a pump will destratify the tank. The third

method of getting heat from the drainback tank to the storage tank is to have a heat exchanger below the minimum fluid level inside the drainback tank. There are commercially available drainback tanks with a submersed heat exchanger. A second pump is required with this system to circulate water from the storage tank and through the heat exchanger and back to the tank. These systems work very well.

Large commercial water heaters or space heating systems can use the drainback design. These systems use a large combination heat storage/drainback tank. This tank can be of any size and is usually vented. A heat exchanger is placed inside this tank to extract the heat from the tank to the load, whatever it is. The water level inside this tank must be checked periodically because some evaporation can take place. It is important to locate the high-head circulating pump as low along the side of this tank as possible, well below the water level inside the tank. This style of solar heating system is a good choice for large commercial systems that do not have a consistent daily heating load,

FIGURE 5.7
Flow meter

especially in warmer climates, because they can completely shut down without damaging the solar fluid.

Closed-Loop Antifreeze Systems

Closed-loop antifreeze systems are the most versatile of all solar heating systems and the most widely distributed type worldwide. Their flat plate collector arrays can be mounted in almost any imaginable way and located at considerable horizontal distances from the exchanger. They are an excellent choice for all climates except hot climates where there is not a reliable and consistent load every day. They are not necessary in climates that do not experience freezing conditions, but they are the only failsafe system for climates that do experience freezing conditions, and should be the only choice for those climates that experience prolonged or severe cold weather or for climates that experience heavy snowfalls.

FIGURE 5.8

Drainback system with tank heat exchanger

Closed-loop antifreeze systems are indirect active systems. With the exception of closed-loop thermosiphon systems, they always use a pump or pumps to circulate fluids through the system.

Closed-loop antifreeze systems are very similar to drainback systems in many ways, but also differ in a few significant ways. These antifreeze systems use flat plate or evacuated tube collectors that heat a solar fluid, usually a high-temperature propylene glycol-water mixture, which is circulated through them. Two insulated copper pipes connect the collectors to a heat exchanger. A relatively small circulating pump is installed on the pipe that feeds the collector array, along with

FIGURE 5.9

PV-powered closed-loop antifreeze system

an expansion tank, some drain valves, a safety valve, and a gauge. A loop of piping starts at the heat exchanger, travels through the collector array, and then completes the loop traveling back to the heat exchanger. This loop of piping is completely sealed and is kept completely full of the solar fluid at all times. These systems can be controlled and powered by a PV panel and a DC pump or they can use a differential temperature controller operating a 120 volt AC pump. When the pump starts, the solar fluid is circulated through the collector array, through the hot supply pipe to the heat exchanger, and then back to the collectors through the return line. When the pump stops, the solar fluid simply stops moving within the closed loop. Because the solar fluid stays in the whole closed loop, there is fluid in the collector array and the piping which travels through unconditioned space. The solar fluid must be able to not only act as a heat transfer fluid, it must also protect the system from freezing. Please study the previous section that describes the characteristics of antifreeze. Methods for transfer-ring the heat from the solar fluid to the domestic water will be described below.

The major limitation of closed-loop antifreeze systems is the tendency of the solar fluid to degrade under high temperatures. The best antifreeze solution today is high-temperature-formulated propylene glycol. This fluid will eventually break down and form a corrosive solution that can harm system components. High temperatures degrade the fluid, and the rate of degradation is directly proportional to the intensity of the overheating, so the hotter it gets, the more quickly the fluid will degrade. All overheating scenarios can be avoided, so this should not be considered a fatal flaw. The best way to reduce overheating is to make sure the system circulates fluid whenever it is sunny out. It's very important, if you have this type of system, to check the condition of the solar fluid regularly. Propylene glycol solar fluid solutions need to be changed periodically. Think of it like changing the oil in your car or truck, except that you only have to do it once every 10 to 15 years under normal conditions.

Closed-loop antifreeze systems are used for residential water heating systems and residential space heating systems. They are also used for commercial water heating systems and space heating systems. Most manufacturers offer antifreeze system kits that are pre-engineered to take the guesswork out of designing a system. They work great for the majority of these installations. Since most solar water heating systems are the same size and installed the same way, these kits are fine for most situations. I highly recommend these kits if purchased from a reputable manufacturer. System sizing is covered in Chapter 7, but you should know that there are only a handful of system sizes that will satisfy the majority of domestic water heating demands, so system sizes and system components are fairly uniform.

To make sure that closed-loop antifreeze systems function properly, a few rules must be adhered to:

➤ The hot supply pipe between the collectors and the heat exchanger must always be made of copper and must be insulated with high-temperature pipe insulation because under rare occasions it can get very hot for short periods of time. Copper is the only practical material that can withstand this heat.

➤ On residential solar water heating systems, the return line from the exchanger to the collectors should also be copper, but standard high quality pipe insulation can be used on that line.

➤ In large solar space heating systems the return line can be made of Pex tubing, except for the final ten feet where it attaches to the collectors.

➤ Collector arrays should be mounted with a slight slope towards the return inlet where the solar fluid enters the array and all piping should slope slightly toward the heat exchanger. For arrays that are remote a drain should be placed at the inlet where the solar fluid enters the array to facilitate drainage.

➤ All horizontal pipe runs should be supported every five to six feet and all vertical runs should be supported every ten feet.

Large space heating systems and large commercial systems are often closed-loop. These systems can be very large and may be the only option for any commercial application where the collectors cannot be located above the storage tank and heat exchanger. A shunt load must be provided on large systems to reduce overheating during non-load conditions. A complete discussion of these types of systems is included in Chapter 6. The world's largest flat plate water heating system, in Green Bay, Wisconsin, utilized over 5,250 – 4' x 8' flat plate collectors (157,689 square feet of collector area). This was a closed-loop solar water heating system and it operated and performed extremely well. That system produced over 37,500 million Btus annually. This system was taken out of service when natural gas prices reached very low prices during the late 1980s. Now that natural gas prices are higher than ever, they wish they hadn't decommissioned the system.

Large space heating systems, seasonal systems, and systems that may experience periodic idle times with no load present, will require a heat diversion load, sometimes called the shunt load. This shunt diversion load helps keep the solar fluid from overheating. A shunt loop is typically a buried length of uninsulated pipe or a radiator located outside, preferably in a cool or windy spot. Outdoor hot tubs are also a common shunt load.

THERMOSIPHONS

Because fluid remains in the solar loop at all times, a method to prevent thermosiphoning must be included in the solar loop. Thermosiphoning can happen when the collectors are colder than the storage tank, but only if the collectors are located above the storage tank, as with roof-mounted arrays. Heat from the storage tank and heat exchanger could rise up the supply pipe to the top of the array and cold heavy solar fluid could drop down the return pipe to the heat exchanger from the bottom of the array. Under this scenario the fluid is circulating in the opposite direction than when it is heating. A check valve will stop this from happening. If a check valve is used in the system, it should be located between the two boiler drains that are used to charge the system. The check valve will also make sure the fluid flows correctly when charging the system. (Check valves were more fully described in the previous chapter.) They are only required in the solar loop when the collectors are above the heat exchanger and storage tank. If they are below, reverse thermosiphoning will not take place. If no check valve is used in the closed loop, then a full-port ball valve must be installed between the drains to facilitate charging. There are occasions, however, when reverse thermosiphoning is a good thing. For instance, if you are going on an extended vacation (for over two weeks) you may want to keep your tank cool to prevent over-heating from speeding the degradation of the solar fluid. You can install a vacation bypass around the check valve to allow for reverse thermosiphoning to occur. This method would not be effective if the collectors were located below the storage tank.

EXPANSION TANKS, PRESSURE, AND PUMPS

The solar fluid within the solar loop is put under pressure when the system is charged. The pressure is typically set at 32 pounds per square inch at 60°F fluid temperature. When the fluid is heated, it expands. As it expands the pressure within the closed loop rises. Eventually, the pressure would rise enough to burst the piping. To prevent such a catastrophic event, an expansion tank is installed on the closed loop. (Expansion tanks are fully described in the previous chapter.) When the fluid cools during the night, or especially during the winter in cold climates, the pressure will drop because the fluid contracts. The expansion tank will provide some compensation for this situation too. It is not unusual to see the pressure in the solar loop fluctuate between 10 and 45 pounds per square inch seasonally, and sometimes daily. Set the pressure of the expansion tank three to five pounds per square inch below the pressure in the system at 60°F, as detailed in the previous chapter. The best location for the expansion tank within the closed loop is directly before the circulating pump in relation to fluid flow. This maximizes flow by compensating for the negative pressure that is created on the

FIGURE 5.10
Vacation bypass

suction side of the circulating pump. Because the expansion tank compensates for pressure changes within the closed loop, it is the point of no pressure change within the closed loop, and it eliminates the negative pressure or vacuum caused by the pump. It is not absolutely necessary to follow this suggestion, but your system will have optimum circulation if you do. The expansion tank should always hang below the pipe it is attached to. Heat will accelerate the deterioration of the bladder inside the tank, and by hanging it below the pipe, the tank will stay cooler.

A pressure gauge is located in the closed loop to monitor the pressure within the loop. This gauge should hang down from the pipe and be visible from the charging drain valves. By hanging below the pipe, no air will get trapped in the fitting or gauge. Unfortunately, most pressure gauges are set up to be above the pipe, so when you install one hanging below the pipe, the scale on the gauge will be upside down. Don't panic, you will get used to it.

A pressure relief valve is also fitted to the closed loop. Note that this is a pressure-only relief valve. The relief valve should be set to open at 80–90 pounds per square inch. Be sure to install a drainpipe on the relief valve and terminate the drain near the floor.

The solar loop circulating pump has no static head pressure to overcome because all the pipes are filled with fluid at all times. The only head pressure the pump has to overcome is friction head. Therefore, low-head circulating pumps can be used to circulate the solar fluid through the solar loop. PV-powered pumps can be used in these systems because the pump can start slowly and circulation will start immediately. In fact, a PV-powered system is best for several reasons. Most importantly, PV pumps run whenever the sun is shining, so there is always circulation in the solar loop when the collectors could be hot. This extends the life of the solar fluid by eliminating stagnation. The only time temperatures over 180°F are encountered in these systems with PV power would be in residential water heating systems during no-load times like vacations during the summer. Dangerous conditions are only possible in hot climates under this scenario and a drainback system is probably preferable where this would occur. PV-powered pumps naturally run at variable speeds: when the sun's resource is low, the pump runs more slowly. The thermal collector's output is also lower under this condition, so pump speed is perfectly matched to thermal collector output. On the other side of the coin, under full sun conditions, the pump runs faster and matches the temperature of the thermal collector by increasing the flow.

A differential temperature controller can also control closed-loop systems. When this is the case, a low-head 120 volt AC pump is used in the solar loop. A complete description of how these controllers operate is included in the previous chapter. If a differential temperature controller is used, I suggest disabling its high-limit function to assure that the pump will circulate at all times that the collector is hot. If

FIGURE 5.11

AC powered closed-loop antifreeze system

the power goes out during a sunny period and a controller is used to regulate the pump, the pump will stop and stagnation can occur. If this occurs for a long time, deterioration of the solar fluid may occur, so make a note to yourself to check the condition of the solar fluid more often.

VENTING AIR

When filling the system with solar fluid it is essential to get all the air out of the system. As mentioned above, the solar fluid, once installed, will be

in the closed-loop system for at least ten years under normal circumstances. Air can cause circulation problems, so a good job done at the start will result in many years of trouble-free service. (The system charging process is detailed in Chapter 9.) Many installers over the years have specified placing an automatic air vent at the highest point in the closed loop. I do not recommend this practice. Automatic air vents fail on a regular basis and should not be used on closed-loop systems. If you faithfully follow the charging process that is outlined in Chapter 9, you will get all the air out of the closed loop. If you still feel that you require a

FIGURE 5.12

Schrader valve air release

Solar Thermal Panel

Schrader Valve

Ball Valve

port to bleed air from the system, install a short riser at the high point of your system on the collectors. Install a ball valve on that riser and install a Schrader valve on the end of the riser. Air will accumulate in the riser because it is the high point in the loop. You can open the ball valve, open the Schrader valve to expel any accumulated air, and then close the ball valve.

Some installers who are used to installing traditional hydronic heating systems will want to install an automatic fill valve on a solar closed-loop system. They do this out of habit because most hydronic heating systems require this valve to keep their system pressurized. Never install an automatic fill valve on a closed-loop system. It will destroy the system by diluting the solar fluid. Diluted fluid can lead to decreased freeze protection.

An automatic air vent must **never** be installed on the solar loop! It will fail and is not necessary if you charge the system properly in the first place. If you feel that you have to have a means of bleeding air out of the solar loop, install a tee at the solar loop's highest point and then add a short length of pipe pointing up with a ball valve in that pipe and a plug in its end. Because this is at the highest point in the system,

any air in the solar loop will accumulate in the pipe and can be manually vented using the ball valve. When you are done bleeding the system, make sure to seal the vent port tightly.

HEAT EXCHANGERS

Closed-loop systems can use either a tank-integrated heat exchanger or an external heat exchanger. Tank-integrated heat exchanger systems are the simplest to install and are suited for situations where space is very limited. These systems are always single-pumped systems, as the only pump required is the solar loop pump. External heat exchanger systems are more efficient and can also be a single-pumped system if the heat exchanger is a thermosiphon type. A plate-type heat exchanger or a coiled tube-in-shell heat exchanger can also be used. In these cases, the system must be double-pumped. Double-pumped systems can be either PV-powered or use a controller and two 120 volt AC pumps. Remember to calculate the correct wire size when running long low-voltage wires between the PV panel and the DC pumps. A complete discussion of heat exchanger types is included in the previous chapter.

Refrigerant Solar Water Heaters

Heat pumps work on the principle of latent heat, by which a fluid absorbs energy (heat) when the liquid vaporizes. When the fluid condenses, heat is released. In a solar energy system, a fluid called Freon absorbs heat in the collector as it evaporates and releases that heat in the solar storage tank.

Because heat pumps can work at very low temperatures, they can extract heat from low quality heat sources. For instance, during cloudy weather, when most liquid solar collectors will perform poorly, a refrigerant system can concentrate the low-grade heat for water heating. Some manufacturers of refrigerant solar water heaters have claimed they don't even need the sun to work: their systems absorb heat from the surrounding air. When operated like this, they are very similar to the standard home heating heat pumps that are popular today in moderate climates.

Most heat pumps power a compressor with an electric motor that can vary in size from a small one-eighth horsepower motor to well over several horse power, depending on the amount of heat that needs to be moved. Heat pumps are rated by their Coefficient of Performance (COP), a ratio of the energy moved divided by the energy it took to accomplish the task. Most heat pumps can deliver around three times the energy required to operate the system. Put another way, a heat pump requires about one kilowatt-hour of electricity (3,400 Btu) to move 10,000 to 13,000 Btu. This equals a COP of 2.5 to 3. An electric heater at 100 percent efficiency has a COP of 1 because it delivers one Btu of energy for each Btu it uses.

So this sounds pretty good, doesn't it? Might there be something wrong with this picture? Let's look a little deeper. First of all, electricity is expensive. The high operating costs of refrigerant-based solar water heaters compared to the operating costs of other solar water heaters has been the main reason this technology hasn't gained in popularity. Second, this method of using the COP to rate the system misses a crucial calculation. Electrical generating and transmission losses associated with conventional fossil-fueled power plants equals about 66 percent. Therefore, if we factor in the losses of generating and delivering electrical energy, the COP of a heat pump system is really only 1, and is similar in efficiency to a gas water heater. Third, as mentioned above, most heat pump systems use Freon as the refrigerant. When released into the atmosphere, many types of Freon reduce the ozone layer that protects the earth from harmful solar radiation.

Heat pump solar heating systems have been promoted by electrical utilities. They use the initial COP rating to classify these systems as being energy efficient, and compared to resistance heating, they are. But as we have seen, they are no more efficient than burning gas. A few companies continue to sell heat pump based solar energy systems. In reality, these systems should be classified as solar-assisted heat pumps because they are really just old-fashioned heat pumps that have been given a slight boost by solar input.

Solar Heated Pools

Two kinds of pools can be solar heated: seasonal swimming pools and indoor, year-round swimming pools. We will also briefly discuss solar hot tub and spa heating systems.

SEASONAL POOLS

There are more square feet of solar swimming pool heating collectors installed per year than all other solar thermal collectors combined. In fact,

18 pool collectors are sold in the US for every flat plate or evacuated tube collector. These are very popular solar heating systems because they are so cost-effective. These systems utilize unglazed collectors made of a special polypropylene plastic. The piping is traditionally PVC pipe. They are usually draindown systems and therefore are direct systems. These systems typically consist of the solar collectors, the piping, a controller and a diverter valve. The pool filter pump usually circulates the fluid through the collectors, so no additional pump is usually required.

Solar swimming pool heating systems can eliminate the need for fossil fueled heating systems. A solar pool heating system can raise the temperature of your pool up to 15°F above ambient temperature. If you use a pool cover, which I highly recommend, you can raise the temperature by another ten degrees. Most solar pool heating systems will pay for themselves in two or three years as compared to using fossil fuels.

The solar pool heating controller has two sensors, one to measure the temperature of the pool water and one to measure if solar energy is available. The controller usually has a dial that sets the desired pool temperature. Most controllers work in conjunction with your filter timer which is set to operate during the daytime. When solar energy is available and the pool needs heat, the controller turns on a

motorized diverter valve, also called an actuator, so the pool water is directed to the solar collectors. As the pool water flows through the collectors it gains heat, usually about one degree Fahrenheit with each pass. After exiting the collectors the water then returns to the pool. The diverter valve is located near the exit of the pool fil-

FIGURE 5.13

Seasonal pool system

ter, so the water is filtered before it enters the collector loop. Pool filter pumps are usually large enough to easily pump the water through the collector loop.

The diverter valve is a three-way valve. When solar energy isn't available, or when the pool reaches its desired temperature, the controller returns the diverter valve to its normal position. In this position water flows straight through the valve and goes from the filter directly to the pool. In the solar heating position, the valve diverts the filtered water to the collector array. When the solar heating system turns off, all the water that was in the collector array and the solar loop piping drains down into the pool. In order to facilitate complete drainage, a vacuum breaker is typically installed at the bottom end of a collector array on the opposite end to where the water enters.

As we mentioned in Chapter 3, plastic collectors are preferable for pool heating applications because most pool water is treated with chlorine, which is not compatible with copper. The collectors can be unglazed because high temperatures are never required, and because they are only used when the ambient outdoor temperature is above 50°F. Each collector manufacturer has its own method of fastening their collectors to a roof or mounting structure. These mounting systems typically consist of special strapping to hold the collectors tight to a roof or flat surface. Please refer to the instructions that come with your system, as they all vary slightly. As mentioned in the collector descriptions, we like the web and tube collector configuration best. This allows the collectors to expand and contract without kinking or binding; in fact, they move around very little. This configuration is also easier to repair than other types. These collectors will last 15 to 20 years, and most manufacturers offer long warranty periods.

Follow manufacturer recommendations for collector array sizing, layout options, and pipe sizing recommendations. For commercial or large residential pools, be sure to contact the manufacturer for their recommendations. Collector array sizing is based on climate, pool surface area, your desired water temperature, amount of shading on the pool, orientation of your collectors, and use of a pool cover. In general, the size of your collector array will equal between 50 percent and 100 percent of the pool's surface area.

INDOOR SOLAR SWIMMING POOL HEATING SYSTEMS

Indoor swimming pools can be heated year-round with solar energy. In most cases if the solar heating system is designed to provide around 50 percent of the heat during the winter months, it will provide around 100 percent during the summer months without overheating the pool. These systems are very cost-effective because they are used to their capacity every sunny day of the year. Indoor pools have exactly the same heat load all year round. The reason the systems provide less heat during the winter is because the solar resource is less then.

These systems are typically closed-loop antifreeze systems. They are installed just like solar water heating systems and solar space heating systems with the exception of the heat exchanger. Special heat exchangers are required because pool water is corrosive to copper. Large swimming pools have surge tanks to hold excess water from the pool when many people jump in the pool at the same time. These surge tanks also usually contain the filters. Pex tubing can be coiled inside the surge tank to act as the heat exchanger. For pools that do not have a surge tank, the heat exchanger is located in the filter piping directly after the filter. With closed-loop systems you cannot just turn off the collector loop like you can with seasonal pool heating systems when the pool reaches your comfort temperature. The collectors will be on every sunny day, so sizing is more critical to avoid overheating the pool.

These systems can be PV-powered. Only one pump is required — the solar loop pump. For pools with a surge tank, that is the only control needed. For pools without a surge tank the pump filtering system and pump must be set to run during the daytime hours. Because the year-round solar resource varies widely from one climate to another, I always use a computer-modeling program like Retscreen to help with array sizing.

The payback on indoor year-round solar heating systems is a little longer than for seasonal pools because the equipment is more expensive and the systems must be slightly under-sized to prevent overheating during the summer months. Mounting the collectors at an angle to maximize the winter sun's resource can reduce the probability of overheating. Sizing is calculated by considering the pool's surface area, the rated collector output, the desired pool temperature, and the use of a pool cover.

Solar Heated Hot Tubs and Spas

Hot tubs and spas are good candidates for solar heating. This is especially true if you already have, or anticipate getting, a solar water heating system or a solar space heating system. The kind of hot tub or spa you have will dictate the kind of solar heating system that is appropriate for your application. The biggest cost in heating these items comes from first filling them and warming the water initially. Maintaining the temperature in a well-insulated tub does not take as much energy. It makes a lot of sense to insulate your tub well.

I have a wooden hot tub outside my home on our deck that holds around 600 gallons of water. The tub is part of the shunt load for my solar space heating system. Our tub (actually it is my wife's tub) serves two purposes: it is a place to put the excess heat generated by our solar heating system during the non-heating part of the year, and it also is a luxury item that our whole family thoroughly enjoys. For half the year our tub is 100 percent heated by solar, and it is tempered all year round. The tub has wooden benches set around the inside near the bottom. We placed 100 feet of 1 inch Pex tubing under the benches and we divert the solar fluid through the tubing. This heat exchanger does a nice job of heating the tub and keeps the fluid cooled. Many of the systems we have designed over the years utilize a tub like this as a shunt load. We have been selling and using the Snorkel Hot Tub for many years and highly recommend that brand. People love them!

There are two approaches to using solar energy to heat hot tubs and spas: a dedicated solar heating system that serves the tub, or combined tub/spa heating with solar water heating. Only wood stave (Ben calls them pickle barrel) hot tubs lend themselves to dedicated solar heating systems. One 4' x 8' collector works well to heat and maintain them. A heat exchanger can be added to the tub as described above. A simple closed-loop system is set up. PV power and a DC pump is best. Plastic hot tubs and spas do not lend themselves to dedicated solar heating systems because it is impossible to get a heat exchanger into the water, as the inside is molded plastic. And because these spas are kept hot all the time, there is no place to dump heat when the sun is out without overheating the unit and there is little heat load per hour. Therefore, the only practical method is to heat a reservoir during the day with solar energy and then take heat from the reservoir when needed.

Under normal circumstances, the tub or spa will be associated with a dwelling or business, which will have a domestic water-heating load. I suggest incorporating the tub/spa heating system with a solar water heater. As in swimming pools, spa and tub water is often treated with chemicals that are not compatible with copper. All piping added should be schedule- 40 PVC solid core pipe. Any heat exchanger that comes into contact with tub or spa water should be stainless steel or cuprous nickel. Most plastic tubs or spas are well insulated, so once these are hot, they do not have a large heating load. Usually one 4' x 8' collector is all that is needed to maintain their temperature. By adding one panel to your solar water heating array and adding around 30 gallons of storage capacity to your solar water heater you will be most of the way to completing a system. You will also need a heat exchanger as mentioned above, a small 120 volt AC circulating pump, and two aquastats to get the heat from the solar storage tank to the tub or spa. Install the heat exchanger in the plumbing line directly after the tub or spa filter so the filtered water flows through the heat exchanger before it returns to the tub or spa. Run insulated copper piping from the top of the storage tank to the hot inlet of the heat exchanger and from the cold outlet of the heat exchanger to the bottom of the storage tank. Install the circulating pump on the cold line, pumping from the exchanger to the tank. Wire

FIGURE 5.14

Pickle barrel hot tub

Benjamin Nusz

the system by tapping into the 120 volt AC line coming out of the timer on the tub or spa that runs the filter pump. Run that line to the first aquastat, then to the second aquastat, and then to the circulating pump. Set the first aquastat to turn on at temperature drop and set the temperature to 5°F above the setting on the tub or spa heater. Set the differential at 3°F. Put the sensor for this aquastat so that it measures the water temperature. Set the second aquastat to turn on at temperature rise and set the temperature at 5°F above the tub or spa temperature and set the differential at 3°F. Place the sensor of the second aquastat to measure the temperature of the tank about one third of the way down from the top of the tank. Whenever the tub or spa filter is engaged, the second aquastat will see if there is heat in the storage tank to give to the tub/spa, and the first aquastat will make sure the tub or spa will not get too hot from the solar heating system.

Chapter 6

SOLAR SPACE
HEATING SYSTEMS

In MANY PARTS OF THE WORLD, the average household spends a significant portion of its annual energy bill on heating. In our quest to use less fossil fuel, we can look to the sun to help reduce our dependence on fossil fuels.

This chapter deals with active solar space heating systems. An active system is one in which we utilize a solar collector to capture the sun's energy. We will not deal with passive solar heating in this book. Passive solar heating utilizes the building itself as the solar collector.

This chapter will outline the available types of active solar heating systems and will detail the systems that have proven to work the best and are the most practical. The type of system you might consider will depend on the kind of building you have, how that building is used, and the climate it is located in.

Unfortunately, most buildings were built to minimum energy standards and that trend continues with new construction. Whether you are building new or retrofitting an existing building, you will get the best return on your investment by making energy conservation the first step in reducing your energy bills. It makes economic sense to do whatever is possible first to increase your building's energy performance and reduce its energy demand, and then size a solar heating system to address that reduced heating load.

It is very rare that a solar heating system can provide 100 percent of a building's annual space heating requirements. There are usually many cloudy days during the heating season and it is impractical to have a heat storage mass that can hold enough heat to do the job. In a high-performance building, though, it is possible to provide a majority of the annual heating with solar energy, and in an average building a significant amount is possible. How far can we go with solar heating systems? For all systems except high-mass radiant systems, I would say something just around 50 percent of the yearly total space-heating requirement is a practical goal. High-mass systems can provide most of the annual heating load for a building, even in very cold climates.

Active solar space heating systems can be classified as those with heat storage and those without storage. When we spoke about solar water heaters, we addressed the fact that the sun only shines directly into our collector for three to eight hours a day. If we want solar heated water to use after the sun goes down, we must incorporate some way of storing the heat collected during the day for later use. We also acknowledged that the sun does not shine every day, so we want to store that solar heat for as long as is practical. Some climates, like mine here in central Wisconsin, can experience a whole week of cloudy weather during the early winter months. During the whole month of December, we may only have five or six sunny days.

Another variable that needs to be kept in mind when sizing solar heating systems is the fact that the solar energy system's output is variable. Likewise, the heating demand of the building will vary greatly over the whole heating season. If we size the solar heating system to heat the building 100 percent on the coldest day of the winter, then on sunny days that are not as cold we will have excess heat and could overheat something. If we size the solar heating system to heat the building on an average sunny winter day, during the worst weather we will not get all our heat from the solar heating system, but over the whole length of the heating season we could get a significant amount.

Solar space heating systems can also be effective without storage. I call this type of system a dump system because heat is delivered directly from the solar collector to the heated space; it is dumped there for immediate use. In most instances, the interior of the building and its contents actually act as a heat storage medium, so even in this type of system there is some minimum amount of storage.

The two most popular types of collectors to use for space heating are air-type collectors and liquid-type collectors. These types of collectors have been described in Chapter 3. Both have been used with success for solar space heating. Each has its advantages and disadvantages and best applications.

Liquid-Type Solar Heating Systems

Liquid-type solar heating systems gained popularity during the early 1980s and continue to be a popular option today. Several new innovations since the 80s make this an even more attractive and practical choice today. We can break this type of system into two basic types: systems with storage and systems without storage.

As was mentioned in the previous chapter, solar water heating systems are the most cost-effective solar energy system. Solar heating systems are the next most cost-effective. If a building

has a domestic hot water load, your first choice in solar energy should be a solar water heating system. If you want to use solar energy to heat your building and you will have a solar water heater, the most practical choice is a combination solar water heating and space heating system. Whenever a solar thermal system is a combination system, the collectors are to be tilted to maximize the winter sun.

LIQUID-TYPE SYSTEMS WITHOUT STORAGE

I call systems without storage dump systems. In this case a liquid is heated in the solar collector and is plumbed into the building. The hot liquid is then directed to some type of liquid-to-air heat exchanger. Familiar types of liquid-to-air heat exchangers include baseboard heating units, old-fashioned radiators, radiant floor tubing, and fan convectors. Details about heat delivery methods are given later in this chapter. If this is a combination system, the excess heat generated by the collectors can be dumped into the building for space heating.

Systems without storage only add heat to a building when the sun is shining and the system is turned on or in the heating mode. These systems should be sized to heat the building on an average winter day, where the output of the collector array is matched to the heat load of the building. If the collector array is too large, the building could suffer from overheating or the system itself could overheat. The solar collectors should be oriented to maximize the low winter sun angle so their output will be the highest during the heating season and decrease in the spring and fall. These systems do not have to work at all during the non-heating season, so the angle of the collectors can be very steep or vertical. This orientation will minimize the potential to overheat the solar fluid during the non-heating season and will help match the collector output to the heating load of the building.

DRAINBACK SOLAR HEATING SYSTEMS WITHOUT STORAGE

The solar loop for a drainback solar heating system without storage is exactly like the solar loop used in a drainback solar water heating system. The best kind of drainback tank to use with these systems is the kind that has a heat exchanger inside the tank. Whenever heat is available, the secondary pump circulates water through the heat exchanger and through the

heat delivery system, dumping heat into the building.

The advantage of this type of dump system is that once the building gets up to its high limit, the system can simply turn off. Also, during the non-heating season the system simply stays off and there is no concern about overheating the solar fluid. The disadvantage of this type of dump system is that it should not be used in cold climates where extended or severe cold conditions are expected.

For a combination drainback solar water heating and space heating system, the solar loop and the storage loop are configured exactly like a plain drainback solar water heater except that the collector array is increased in size to accommodate the heating load. Of course, all the piping and drainback tank size would be increased proportionately to account for the increased array size. This configuration will heat the storage tank rather quickly because the collector array is oversized for the storage tank (which is sized for the domestic load only). An aquastat is added to the storage tank and set at 130°F. When the storage tank gets above 130° a pump comes on and either delivers hot water from the storage tank to the heat delivery equipment or it delivers heat to a heat exchanger where the heat is transferred to another liquid that goes to the heat delivery equipment. Note that some plumbing codes allow domestic hot water to be used in a heating system as long as the piping is approved for potable water and as long as there is no chance that the water will be contaminated by another liquid. In other words, you could not send domestic hot water into a hydronic heating system that was heated by another source. Some codes never allow domestic hot water to be used for space heating, so in those areas, the second heat exchanger is required.

A thermostat is placed in the line between the aquastat and the heat-circulating pump and is set to open on temperature rise. This thermometer is used to control overheating and should be set at the maximum allowable temperature. When the building reaches that temperature, the thermostat will turn off the pump.

CLOSED-LOOP ANTIFREEZE HEATING SYSTEMS WITHOUT STORAGE

Closed-loop antifreeze systems without storage follow the same description given above for the drainback systems with the following exceptions. If the antifreeze system is not a combination system, the solar fluid can be sent directly to the heat delivery equipment. Antifreeze systems cannot just be turned off like the drainback system can, so care must be taken to keep the antifreeze from overheating. Adjusting the tilt angle of the collectors will help, but oversizing the collector array can be a problem, especially in warm climates where the heating demand is small during the winter. I also suggest not using the high-limit thermostat in the building, and just letting the system run. If it gets too hot, open a window. A diversion load should also be supplied to shunt excess heat during the summer to keep the solar fluid from overheating. Common diversion loads include buried pipes or radiators placed outside the building. The operation of diversion loops is described in the high-mass section below.

Liquid-Type Systems With Storage

Liquid-type solar heating systems with storage are simply expanded solar water heating systems. There are two kinds of heat storage mediums that are used with liquid type systems: water and sand.

DRAINBACK SOLAR HEATING SYSTEMS WITH STORAGE

The solar loop for a drainback solar heating system with storage is very similar to a drainback solar water heating system solar loop. There are two variations: one uses a single tank that acts as a drainback tank and heat storage tank; the other uses a conventional drainback tank and several conventional steel heat storage tanks with internal or wraparound heat exchangers.

For systems with a large single storage tank, the water from the storage tank is pumped through the collectors when solar energy is available. The high-head pump is located as low as possible next to the tank and below the water level inside the tank. When the system turns off, the water that is in the collectors and piping drops back into the large tank. The system uses a differential temperature controller to activate the pump. Whenever solar energy is available, the large tank gets heated. There is a high-limit setting on the controller, so when the tank reaches its desired temperature, the system turns off and the water drains from the system back to the storage tank. A heat exchanger is submerged in the tank to extract heat for the heating system. A combination system would have a sec-

ond heat exchanger in the storage tank to heat the domestic hot water. Details about heat storage configurations and operation are given later in this chapter.

This configuration's main limitation is the likelihood of its freezing if installed in a climate with prolonged or severe cold conditions or large amounts of snow annually. Another limitation is the distance from the collectors to the storage tank. A large enough circulation pump must be used to overcome the static head from the level of the storage tank to the top of the collector array. This could be a serious issue if the building is more than two stories tall. Like all drainback systems, the collectors must be located above the storage tank to facilitate proper drainage. This eliminates the possibility of remote-mounting the collector array.

The big advantage of these systems is their ability to simply turn off and drain when the storage tank reaches its high limit. This is especially advantageous where the load is not consistent over the course of a week or in warm climates.

For systems that will use multiple storage tanks, a conventional drainback tank is used and water from the drainback tank is circulated through the tank-integrated heat exchangers on each individual tank. Under this scenario the water held in the storage tanks is pumped directly to the heating system. If this is a combination system the water in the storage tank is either used for both purposes or an additional heat exchanger is used for the domestic hot water heating. If a heat exchanger is used, there will have to be an additional circulating pump and control to circulate hot water from the storage tank through the heat exchanger.

This design does not suffer from the height issues that limit the large tank systems because the drainback tank can be mounted a considerable distance above the storage tanks. However they do need to have the collectors above the drainback tank and are limited by the same climatic conditions as all other drainback systems.

CLOSED-LOOP SOLAR HEATING SYSTEMS WITH STORAGE

There are three storage options for closed-loop solar heating systems with storage: systems that use one large storage tank, systems that use multiple storage tanks, and systems that use sand as the storage medium.

The solar loop for all three storage methods is exactly the same and follows the traditional closed-loop design that was described in the solar water heating section.

The traditional method of storing heat for these systems uses a single large storage tank. The solar loop delivers heat to the tank through a heat exchanger located in the storage tank. Heat is extracted from the storage tank through other heat exchangers inside the tank. Multiple heat exchangers can be placed inside the tank, so combination systems are simple to install. The solar loop for these systems can be PV-powered or can use differential temperature controllers. This is the preferred system for retrofit applications and is compatible with all existing heat delivery systems. A diversion load is required to shed excess heat that may be collected during the non-heating season.

This system's big advantages include versatility of collector placement, absolute freeze protection for all climates, and regulated temperature control inside the building. The disadvantage is the system's tendency to overheat when installed in warm climates, especially when the load is intermittent.

When multiple storage tanks are used, the water in the storage tanks is usually sent directly to the heat delivery system if possible, but additional heat exchangers can be used to extract the heat from the tanks. The easiest tanks to install and engineer are those with integrated heat exchangers. Combination systems require a separate storage tank for the domestic hot water. Personally, I am not a fan of these systems, especially if they are combination systems, because they get quite complicated and expensive. The advantages and disadvantages for this style are the same as those listed for the closed-loop system with large single storage tank.

Closed-loop solar heating systems can use sand as the heat storage medium. We will discuss this kind of system in detail later in this chapter. These systems have a traditional solar loop, and like the closed-loop systems with large storage tanks, the solar loop extends into the thermal mass sand bed to deliver the heat from the collectors directly to the storage medium. Combination systems are typical, especially in residential applications. Combination systems have an additional solar hot water storage tank, configured exactly like those in typical closed-loop solar water heating systems. In this case, as the hot solar fluid is brought into the building, some of the hot solar fluid is sent to a liquid-to-liquid heat exchanger to heat the water in the storage tank, while the rest is sent to the storage mass.

These systems differ from all other solar heating systems with storage because their amount of thermal mass is much larger. In fact, they typically have over 50 times the weight. This very large thermal mass has the ability to store an appreciable amount of heat that tempers a building in a significant way. The advantages of this type of system are the same as those listed for the other closed-loop systems, with the added advantage of not having a large storage tank within the building. No storage tank means no tank maintenance or possibilities of leaks. These systems can provide a majority of the annual heating load, even in extremely cold climates. While overheating is always a possibility with any closed-loop system, it is less likely with a high-mass system like this, if properly designed, installed, and operated.

Heat Storage Configurations

In this section we will be describing two kinds of storage configurations, large single tanks and multiple smaller tanks plumbed together to create a volume of liquid equal to one large tank. The large single tanks are almost always open tanks, which means their tops can come off and are not sealed to be a pressure vessel. Both large storage tanks and multiple smaller tank systems can be configured for closed-loop anti-freeze systems and drainback systems.

LARGE STORAGE TANK CONFIGURATIONS

The best solar storage tank is seamless and jointless. With no seams or joints, there will be less chance of a leak as the tank ages. The best tanks will also be constructed of a material that can withstand consistent 180°F tempera-tures. The tank will also have to be insulated and it must have a tight-fitting lid that does not drip. Another important consideration, especially in retrofits, is that the tank must be able to get into the building.

I have seen tanks constructed of almost every conceivable material that can hold water. Many site-built tanks were wooden frames lined with plastic or rubber roofing materials. None of these tanks lasted more than 10 or 15 years — and some as few as 2 years — without developing leaks. I have seen 250 gallon or larger steel oil drums used as solar storage tanks. These were never made to contain water, so they developed leaks over time as well.

Being from the Dairy State, I have seen a number of used stainless steel milk bulk tanks used. These have actu-ally worked very well, because the stainless steel does not corrode and they are welded together, eliminating potentially leaky seams.

My favorite solar storage tank is made of high-temperature-rated fiber-glass. There are a number of these tanks on the market and they are reasonably priced at around $2 to $3 per gallon of capacity at the time of printing. These tanks come in a variety of sizes. The 400-gallon size is around 40 to 48 inches in diameter and 4 feet to 6 feet tall. The tank must come in pieces in order to get through door-ways. The tanks are then assembled in place. Some tanks bolt together using a rubber gasket to seal the joint while others are cold-welded together using special epoxy cement. The epoxy-sealed joints are the best.

The most trouble-free tanks are those with no fittings, which are always potential leaks, especially on round tanks. This means that all the piping to and from the tank must come out of the top of the tank. Because most configurations also require at least one heat exchanger inside the tank, you must have access to the inside of the tank through a lid of some sort, and the bigger the better. The lid should have a flange on the inside to prevent condensation from dripping over the side.

Tank heat exchangers are often coils of copper or Pex tubing suspended within the tank. Copper finned tubing can also be used as a heat exchanger. These heat exchangers can be set on a rack inside the tank or they can hang from the top. You can make a rack out of half-inch rigid copper tube and a few fittings to hold the heat exchanger off the bottom and also hold any exchangers that are near the top. Be sure to put caps on the bottom of the legs so they won't wear a hole in the bottom if they move around through expansion and contraction.

Fiberglass tanks usually do not come insulated, so you will have to install insulation. Set the tank on at least two inches of extruded foam. Wrap the tank with reflective insulation and fiberglass batts with the foil on the outside. Use foil tape to make a nice looking job.

CLOSED-LOOP CONFIGURATIONS

The solar loop of a closed-loop solar heating system runs from the collec-

tors directly into the storage tank and connects to a pipe coil heat exchanger that should be situated just above the bottom of the tank. Type K soft copper pipe of the same size as the closed-loop piping is typically used for this application. It then returns back to the collectors. A PV-powered pump is perfect for this simple circulation circuit. Whenever it is sunny the pump circulates hot solar fluid through the collectors, through the exchanger,

FIGURE 6.1

Closed-loop space heating tank

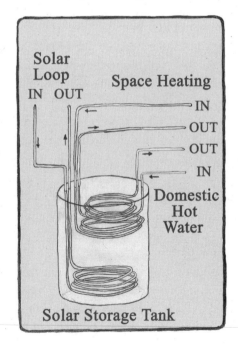

and back to the collectors. This is very effective. A differential temperature controller may be used as well.

Once the tank is hot, you have to extract the heat from the water in the tank or use the water directly for heating. I suggest using another heat exchanger suspended near the top of the tank to extract the heat. All exchangers should be sized using at least one foot of exchanger pipe for each four square feet of collector area and is typically type K soft copper. This is the easiest and most trouble-free method of extracting heat from your tank. You can also use the hot water directly from your storage tank. To do this you need to install a bulkhead fitting on the side of the storage tank a minimum of 16 inches below the water level in the tank. A pipe attached to the bulkhead fitting on the outside of the tank turns down and runs along the side of the tank to near the bottom, makes a 180-degree turn, and then runs in an upward direction for a short way. The heating loop circulating pump is then installed on this pipe, pumping away from the tank. The pump must be located below the water level in the tank to keep it primed. If you have to have a bulkhead fitting on a fiberglass

tank, get a solid brass one, or better yet, have one installed at the factory. Do not overtighten plastic or fiberglass fittings like those sometimes found as bulkhead fittings! We will discuss the ways to operate the heat extraction heat exchanger in the part of this chapter that deals with heat delivery methods.

You can use multiple tanks to store large amounts of heat in water. The best way to heat the tanks is to use integral heat exchangers. You can then plumb the solar loop to pass through all the heat exchangers in the individual tanks. I suggest you plumb the tank heat exchangers in series, with the tank closest to the heat load being the first. The fluid in these two circuits, the solar loop and the extraction loop, will flow in opposite directions. Using multiple tanks will cost around twice as much money as the complete cost of a single fiberglass tank and the fiberglass tank will last longer than individual steel tanks.

DRAINBACK CONFIGURATIONS

Large drainback systems that use a large storage tank are set up just like a drainback solar water heating system. The large water tank stores the heat and is the drainback tank. The tank water is also the solar fluid. A bulkhead fitting is installed on the side of the tank at least 16 inches below the tank water level and a pipe configuration exactly like the one just described above is installed. In this case the solar loop pump is a high-head pump capable of overcoming the static head of the system. The solar fluid leaves the tank and travels down and then up into the high-head pump. It is then pumped up through the collectors where it is heated. After exiting the top of the collector array the heated water then drains back to the large storage tank. The static head is calculated from the water level in the storage tank to the top of the array plus four feet. Heat is extracted from the tank in the same ways as described above for the closed-loop antifreeze system.

You can also use multiple tanks with a drainback system. This system would be set up exactly like a drain-down solar water heater. Use tanks with integral heat exchangers. Locate the drainback tank above the storage tanks and circulate the solar fluid from the drainback tank through the heat exchangers, just like a domestic system except there are multiple tanks and multiple exchangers. Because the drainback tank can be located any dis-

tance above the exchangers, static head can be drastically reduced as compared to the large single tank configuration.

Heat Delivery Methods

There are a number of options for taking the heat out of a water tank and delivering it to the building on demand. Which method you choose will depend on your existing heat delivery system, if there is one, and also might depend on the type of storage tank you have.

INTEGRATING A SOLAR HEATING SYSTEM INTO A FOSSIL FUEL HEATING SYSTEM

Integrating a solar heating system into an existing hydronic heating system is possible and sometimes fairly easy, but at other times it can be difficult and hard to figure out. Because there are so many variations in plumbing styles and methods, it is impossible for me to tell you exactly how to do your installation. If possible, consult with the heating contractor who installed your existing system and seek his advice. If that contractor is no longer around, contact one who is willing to work with you on this project. If you want to do the whole installation yourself, you can learn a lot from studying the information on <www.heatinghelp.com> or other similar sites on the Internet.

If you are installing this system in a new construction, I suggest you use a primary loop-secondary loop plumbing configuration. This configuration allows for multiple heat sources and multiple heat loads in a simply designed system. There are several good books available on this subject.

FIGURE 6.2

Drainback space heating tank

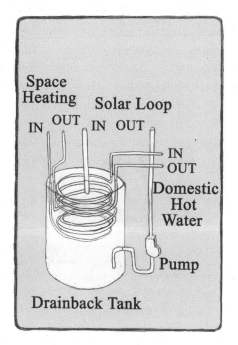

My favorite is "Pumping Away" by Dan Holohan.

A main consideration when integrating a solar heating system with a conventional hydronic heating system is that you do not want to mingle the fluid in your solar storage tank or solar loop with the fluid in your hydronic heating system. The best way to integrate a solar system into a hydronic system is to use a large storage tank with a heat exchanger mounted inside the tank to extract the heat. This exchanger shares fluid with the hydronic heating system (inside the exchanger) so the fluids never mingle. If you are using multiple tanks, you must use an external heat exchanger to transfer the heat from the stored water to the heating fluid. If you are integrating a dump system, you must also use a heat exchanger.

When you integrate a solar heating system into a traditional heating system you have two heat sources to choose from. This can be done automatically by using two thermostats in the building, one for the solar heating system and one for the back-up. Set the solar thermostat slightly higher than the one for the back-up system. If you have multiple zones, you will need two thermostats for each zone.

As the temperature falls in the building, the first thermostat to call for heat will be the solar thermostat, wanting to extract heat from the solar storage system and deliver that heat to the house. The control circuit has to check the solar storage tank to see if any heat is in there to deliver. A sensing device called an aquastat or set-point thermostat does this job. A sensor is immersed in the storage tank and attached to the sensing device. The device can be adjusted to sense the minimum tank temperature that will provide hot enough fluid to work with the heat delivery system. If there is sufficient heat in the storage tank, a pump relay will turn the circulating pump on and heat will be delivered. Pump relays and aquastats are available from any hydronic heating supplier.

Integrating solar heat into existing hydronic systems that operate at lower temperatures, like radiant floor heaing systems, is the best option. Many hydronic heating systems operate at 180°F. This temperature is rarely achieved during the heating season in the storage tank, so integrating solar into these higher-temperature systems is usually not practical.

INTEGRATING A SOLAR HEATING SYSTEM WITH A FORCED-AIR HEATING SYSTEM

It is possible to integrate a solar heating system with a forced-air heating system. This is commonly done in retrofit situations. If you are building a new home, however, I do not recommend a forced-air heating system because their blowers use a lot of electricity to distribute heat throughout a building. People often neglect to include these electrical costs when they compare the efficiency of various heating systems, even though they can add a significant amount to the total cost. For people who already have PV power, or are planning on having PV power, a forced-air heating system is not an alternative because of the large number of watt-hours they consume. The only reason to ever have ductwork and blowers is for heat recovery ventilation systems or fresh-air systems in commercial buildings. There are alternative heating systems with a much lower operating cost than forced-air systems. Although they might cost slightly more to install, their lower operating costs will quickly make up the difference. Ask your heating contractor about radiant floor heating, which is the best. There are many other alternatives as well.

When integrating a solar heating system into a forced-air heating system, the object is to get the heat from the storage tank and into the house. The blower and ductwork associated with a forced-air heating system will do the job, and if they are already in use, they make the obvious choice. A liquid-to-air heat exchanger is placed in the ductwork near the furnace. It is always preferable to install the heating coil in the hot air duct above the furnace. If an air conditioning coil is already installed in the ductwork, it will probably be in the hot air duct just above the furnace. If there is room, try to place the heating coil above it. If there is no air conditioner coil in the hot air duct and if central air conditioning will not be installed in this system, then install the heat exchanger in the hot air duct just above the furnace. Hot water from the solar storage system is circulated through the liquid-to-air heat exchanger and the furnace fan is engaged. As air is moved through the heat exchanger, it is warmed and then delivered throughout the building by the ductwork.

This heating circuit has several control functions that need to be addressed. A low-voltage thermostat (exactly the same component that is used to turn the forced-air furnace on and off) will turn on the solar heating circuit when there is a call for heat, and turn it off when the desired indoor temperature is reached. This thermostat should be set at least several degrees above the setting of the thermostat serving the conventional heating system. This way the solar system will do the maximum percentage of the heating load that it can.

This circuit should only come on if there is sufficient solar heat in the storage tank. Either a set-point thermostat or an aquastat can be used to control this function. These devices measure the temperature of the storage tank and are preset to recognize the lowest temperature we want our heating circuit to operate at. For instance, I usually set the low temperature shut-off point at 85°F for a house where we like a temperature above 70°F. That is a difference of around 15 degrees. With this setting, when the tank temperature falls to below 85°F, the device will turn off the circuit and not allow it to come on.

A circulating pump must come on to circulate the hot water from the storage tank heat exchanger to the heat exchanger in the ductwork and back to the storage tank. This circulating pump is usually high voltage (110 volt AC) whereas the thermostat circuit is low voltage. A relay, which is wired into the thermostat circuit, will do this job.

The furnace fan must also be turned on when the solar heating system is engaged. The furnace has a terminal strip just inside the cover. There will be a row of small screws with letters by each one. These letters will include R, C, and G. The R terminal provides 15 volts AC for the thermostat circuit. A thermostat wire is attached to the R terminal. This wire first goes to the thermostat. Then it goes to the set-point thermostat (or aquastat). From there a branch goes to the pump relay. The other branch goes back to the furnace where it is attached to the G terminal. The G terminal controls the blower. Here is how it all works: the temperature in the building falls to below the desired temperature. The thermostat calls for heat by closing a switch. Low-voltage electricity flows from the furnace (R terminal) through the thermostat and to the set-point thermostat (or aquastat). If the temperature in the storage tank is above our minimum setting, a switch will close in the set-point thermostat and the electricity coming from the thermostat will flow through this switch. Some of this low-voltage power turns on the pump relay, which

turns on the pump. The electrical energy then flows back to the furnace where it goes to the G terminal, which engages the blower.

If there is a central air conditioning system installed on this furnace, a single-pole double-throw switch must also be installed, as a summer-winter mode isolation switch. This is because a central air conditioning system also uses the G terminal to turn on the blower when it is engaged. When either the solar energy system or the air conditioner is turned on, the other system may turn on as well. When installing the switch, connect the solar lead to one side of the switch and connect the air conditioning lead to the other side of the switch. A single wire is attached to the "out" pole of the switch and runs to the G terminal. The switch is flipped one way during the winter to allow the solar heating system to operate, and the other way during the summer to allow the air conditioning system to operate. This switch must be manually flipped at the beginning and end of the heating season.

NON-INTEGRATED HEAT DELIVERY METHODS

When retrofitting a solar heating system into an existing building, it is sometimes impossible to integrate it into the existing heating system. For instance, fan convectors, baseboard radiators, or radiant floor tubing may be used to deliver the solar heat into the building. In these cases the solar heating system will stand alone and work independently of a conventional heating system.

Options for delivering the solar heat into the building include a separate radiant floor system, one or more fan convectors, or baseboard units. With all these options, hot fluid is taken from the solar storage tank and circulated through the heat distribution device where the heat is delivered into the building. The control circuit for any of these options is very simple and the circuit is exactly the same for all of them. This simple control system consists of three basic components: a thermostat, a circulating pump, and a set-point thermostat (or aquastat).

This control circuit will operate just like the circuits described above for the integrated systems. The difference here is that you have the option of the control circuit operating at household voltage. The high-voltage method uses a line-voltage thermostat instead of a low-voltage thermostat. With this option there is no low-voltage power supply and all the wires must be sized for the high voltage. The advantages of this high-voltage system are that it eliminates the need for the low-voltage power supply and the pump relay. The disadvantage is that you have to run bigger wire to the thermostat.

Baseboard units are what you see in homes or buildings with traditional hydronic heating systems. They are composed of a copper pipe with radiating fins located along the length of the pipe. This finned pipe is covered by an attractive metal cover and is installed on the wall just above the floor. As warm fluid is circulated through the pipe, the fins get hot and heat the surrounding air. This warm air rises up along the wall and as it exits the heat exchanger, cool air is pulled in from the floor and is heated. This natural convection results in a very comfortable circulation pattern. Compared to a forced-air system, much less energy is required to distribute the heat throughout the building.

Although they are becoming rare, old-fashioned cast iron radiators can work very well for solar dump-type heating systems. These large units can deliver a large amount of heat into a room efficiently. The downside is that they are big and take up a lot of room.

Fan convectors consist of a liquid-to-air heat exchanger with a fan included to help circulate the heat. These units are often 18 to 24 inches tall, up to 10 inches deep, and can be 36 inches wide. You might recognize these as they are often used in motel rooms and are usually located under the window. Fan convectors are especially useful when installing a solar heating system in an existing building because they are easy to install and do not take up very much room. Plus they can deliver a lot of heat very quickly. Fan convectors can be usefully combined with solar radiant floor heating systems for areas that might be hard to get heat to, like bathrooms, where we like it warmer than other

FIGURE 6.3

Combination space heating system

areas. Fan convectors use more energy to distribute heat into the dwelling than baseboards because they include a blower, which uses electricity.

High-Mass Systems

It is interesting to note that the first known use of radiant floor heating, a high-mass system, was by the Romans over 2,000 years ago. Their famous bathhouses had a central heating system that heated the water used for bathing as well as for heating the building. A firebox was located below the floor level of the bathhouse. The smoke and heat from the fire circulated below the floor, heating it before exiting the building through numerous chimneys. As the smoke and heat circulated beneath the floor, the floor was heated. These Roman bathhouses are known for their popularity and luxuriousness. Way back then, people knew that radiant floors provided the most comfortable heat. Today we know that radiant floor heating is not only the most comfortable, it is also the most economical to operate.

GENERAL DESCRIPTION

A high-mass solar heating system uses a closed-loop antifreeze solar energy system to collect solar energy. This is exactly the same system that is used in a closed-loop solar heating system with storage. Instead of storing the heat in a tank of water, we store it in sand under the floor of our building. The heat then slowly rises up into the building warming the floors and the whole building. If there will be a domestic hot water load in the building, a combination system can be configured so the system will heat the sand and the hot water.

This type of solar heating system uses a large amount of sand located beneath the lowest floor of the building as the heat storage medium. The sand is placed inside an underground insulated box, located under the floor. The insulated box holding the sand is made of two-inch extruded foam insulation.

We get the heat into the sand box using a gridwork of Pex tubing, similar to a traditional radiant floor heating system. The tubing is placed just above the bottom of the box, and hot solar fluid is circulated throughout the tubing, warming the sand.

SAND IS SLOW

The comment I hear most often about high-mass heat storage is that sand is a poor heat storage medium and a poor conductor of heat. It is true that sand stores and conducts heat differently than water, but we can take those differences and use them to our advantage.

When a large amount of thermal mass is used, it can take a long time to heat up. Likewise, it takes a long time for this large thermal mass to cool down. We can use these traits to our advantage. This type of high-mass system can be turned on in the late summer. Because it takes a long time to heat up the mass, we can begin to harvest solar heat well before the beginning of winter without overheating the building. We continue to charge the system with heat throughout the late summer and fall.

This type of system is classified as an unregulated system because there is no thermostat to control the indoor temperature. The fact is that high-mass solar heating systems rarely, if ever, overheat a building. One reason is the collector tilt angle. Collectors are tilted to maximize the low winter sun, so they are mounted at a steep angle. During the warm months of the year the sun is high in the sky and therefore sheds little energy onto the collectors. In other words, the collectors self-regulate to a great extent by being very efficient at collecting heat during the heating season and very inefficient during the non-heating season.

For most locations I recommend a sand bed that extends two feet deep under the whole building. Compacted sand weighs around 105 pounds per cubic foot. With a two-foot bed, our mass weighs 210 pounds per square foot of interior floor space. That equals 10.5 tons per 100 square feet of the building. For a modest 1,200 square foot house, for example, the thermal mass would weigh 126 tons. This is a lot of mass. By comparison, the weight of a 350-gallon tank full of water is only 2,800 pounds (1.4 ton).

Water is the medium to which all other materials are compared in terms of their heat storage capacity. One cubic foot of sand can store the same amount of heat as 4.8 gallons of water. Water is about twice as dense as sand and transfers heat much better because the air spaces surrounding each grain of sand inhibit the flow of heat. But even though sand may not be the perfect medium, it is cheap and easy, so using a lot is not a big deal.

By having a really big thermal mass, we can store a lot of heat. Depending on the overall efficiency of the building housing this solar energy system,

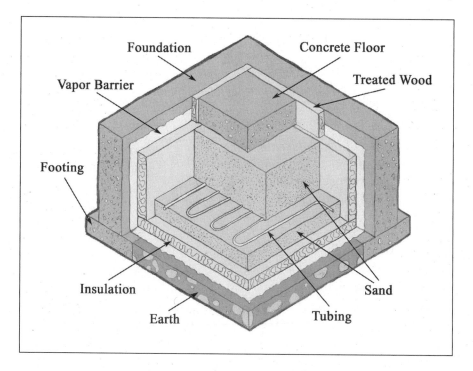

Foundation

Concrete Floor

Vapor Barrier

Treated Wood

Footing

Insulation

Sand

Earth

Tubing

FIGURE 6.4

High-mass system detail

we can usually store several months' worth of heat in our mass. In addition, whenever the sun does shine during the heating season, additional solar heat will flow into the mass, replenishing it.

LIVING WITH A HIGH-MASS SOLAR HEATING SYSTEM

Since every solar energy installation is different, a high-mass solar heating system will react slightly differently in every situation. The first thing to realize is that it may take an entire year to get the sand bed dry and hot. It also makes a big difference when during the year the system is installed. If the solar energy system comes online early in the summer and is allowed to operate most or all of the first summer, that system will perform relatively well the following winter. If the system comes online any other time of the year, the performance of the system will be diminished that first year.

If you look at the section in Chapter 7 about sizing solar heating systems, you find certain general sizing parameters. You will see that for a dump-type system, a certain amount of collector area is needed for a particular building. If we were to add storage to that same building, we would approximately double the amount of collector area. This additional collector area and storage allows us to heat the building for a longer period of time, usually for a whole day instead of a half-day. You will also see that the sizing formulas developed for closed-loop solar heating systems with storage are basically the same as those for high-mass solar heating systems. Lastly, you will see that the high-mass solar heating system might provide 25 percent more heating than the closed-loop system with storage with the same sized collector array. You are probably wondering how this can happen. The reason is that with a closed-loop solar heating system with storage, we turn the system on when the heating season begins in late September and turn it off when the heating season ends in early May. In the case of high-mass solar heating systems, we turn the system on in early to mid-August and operate it through early May. This ear-lier start date allows this high-mass system to collect nearly two additional months of solar energy using the same collector area, thus increasing the overall efficiency of the system. Please note that these additional days of operation come during a traditionally sunny time of year, so we harvest even more solar energy. The large thermal mass allows us to store that energy for later use.

Once it's installed, you will have to develop your own particular operating schedule for your high-mass solar heating system. The earlier in the late summer that you can turn your system on, the better. You will have to experiment. You might find that during the first year or two of operation, you have to turn the system on early in August. After that you might delay the turn-on date till later in August. But if it is cool at night, most people just leave the windows open and cool the building with fresh air. Even with the windows open and the building at a comfortable temperature, the mass is still heating up. The warmer we can get that mass during the sunny months preceding winter, the better. Most people with high-mass solar heating systems are able to keep their windows open much later in the early winter than those using conventional heating. This fresh air is a real bonus for those with high-mass systems, especially in locations like Wisconsin where most buildings are closed up tight for over half of the year.

In the spring you will want to turn the system off once the building is consistently very warm. When you do this will depend on your location and how much sunny weather you are experiencing that particular spring. You might find that during the spring you have the system off for a week and then back on for a week till the temperatures outside become uniformly warm. Of course, the solar water heating portion of the system will continue to operate all year long.

DIVERSION LOADS

All high-mass solar heating systems will have a shunt loop or diversion load. This diversion load is a place where we can send the excess solar heat when we do not want to use it in the building. We want to keep the solar fluid from getting extremely hot, thereby prolonging its life. We do this by sending the collected solar heat somewhere outside of the building. Most shunt loops are a single long loop of Pex tubing, which is buried outside the building. I suggest

one foot of Pex tubing for each square foot of collector area for the shunt. If the soil is extremely dry at your location the shunt won't work as well, so additional shunt length may be necessary.

After the hot solar fluid enters the building from the collectors and then passes through the heat exchanger for the domestic hot water, we divert it through the shunt before it goes back to the collectors. This effectively keeps the solar fluid from getting extremely hot. All solar fluids experience breakdown under overheat conditions, so if we keep the temperatures within tolerances, we will extend the life of the solar fluid and reduce costly maintenance.

To divert the solar fluid from the floor loops to the shunt loop is easy. There is a ball valve on the shunt loop and one on the floor loops. (Ball valves are simple lever-actuated shut-off valves.) Simply close the floor valve and open the shunt valve. This procedure takes 15 seconds to accomplish. **Note: Never have both these valves closed at the same time because then the solar fluid will not circulate, overheating will occur, and damage to the system can be expected.**

Another option for a diversion load is to have a heat exchanger in an outside hot tub. This is the system I use at my home. I have placed a 100-foot-long length of 1 inch Pex tubing coiled around the inside of the hot tub near the bottom. When this diversion loop is engaged, the hot tub stays hot — it's great. I encourage you to be creative when designing the shunt. If you have any uses for some heat during the non-heating months, by all means try and use this excess solar heat. It's free!

Air-Type Solar Heating Systems

One of the most cost-effective solar heating systems is an air-based system. This is because the equipment is simple and inexpensive. Air is the heat transfer medium instead of solar fluid. Ductwork is used to transfer the heat from the collector to the building. These systems are most popular in climates that have large heating loads. On cold, sunny winter days, air systems can produce impressive amounts of heat, especially if there is snow on the ground. Air systems are strictly space-heating systems and are not used for water heating. I will get into that a little later in this section.

Air-type solar heating systems suffer from the same flaw as conventional forced-air systems. A substantial blower is required to distribute the heat collected in the collector through the building. This electrical load can account for up to half of the total electrical demand of an electrically efficient home during the heating season. Air-type collectors are also not as efficient as liquid-type collectors, so more collector area is required. On the other side of the coin, air-type collectors are less expensive to purchase per square foot, so if array size is not a limiting factor, this point is not a deal breaker.

While these systems are very cost-effective, they do have their limitations. The biggest is that they can only be set up as systems without storage. Back in the 1980s, these systems were set up with heat storage components of many kinds. These attempts used rock bins, closets full of bricks, special tubes filled with eutectic salts, and others. Hot air was circulated through these kinds of thermal mass and the mass soaked up the heat and stored it for use at a later time. To get the heat out, cool air was circulated through the mass where it picked up any available heat that was stored there. Most of these systems worked for a while, but problems soon arose. The most significant problem involved air quality. As dust and dampness accumulated on the mass material, a medium was created for the growth of mold and other nasty things. These pollutants then were circulated into the building creating potential health problems. Rock bins also rotted, leaving a huge mess in homes. Eutectic salts, which are very corrosive, leaked out of their containers. Do not attempt to have a dedicated heat storage set-up for an air system.

When set up without storage, air systems can be a good choice for solar heating systems, especially in commercial applications. Two versions of air-type heating systems are used today. The most common type takes cool air from inside the building and circulates it through the collector where the air is heated. The warmed air is then delivered back into the building. Back in the early and mid-1980s, I manufactured and installed many dump-type air systems. Almost all those systems are still working fine today and saving homeowners money. These systems consisted of a wall-mounted collector and minimum ductwork with a blower and controller. They were simple, inexpensive, and were designed to provide around 25 percent of the annual space heating needs.

Most people also enjoyed warmer temperatures in their homes during sunny days. I remember one family in particular. They were a family of five, with both adults being schoolteachers. We installed an air system on their home during the week and were done on Friday. That Saturday about noon I got a call from them. It was a perfect sunny winter day with outside temperatures hovering around 10°F. It seems that these folks were quite conservative with their thermostat settings, so their house was not very warm during the winter months. The call was to invite me to a spontaneous Bermuda shorts party at their house because the solar heating system was making the house warmer than it had ever been during the winter. Needless to say, they were happy.

The less common type of air system uses transpired collectors which draw fresh air through the collector, where the air is heated and then delivered into the building. These systems are designed for buildings that require large amounts of make-up air. Because these systems are for space heating only, they are often wall-mounted. Wall-mounted systems work well during the winter when the sun is low in the sky; they don't work during the summer when the sun is high in the sky. They also work best in buildings that contain a large amount of thermal mass, like warehouses and shops. Because the solar fluid in these systems is air, and ductwork is required to move that solar fluid within the system, most air systems are building-mounted. Running ductwork outside is not very efficient.

AIR SYSTEM CONTROLS AND ADDITIONAL COMPONENTS

It is possible to design PV direct systems that use a PV collector wired directly to a DC blower. This design is most appropriate for small systems that are wall-mounted. As this set-up has no controller or thermostat, there is no overheating control. Also, DC blowers are hard to come by.

I like to use differential controllers with air systems. I put one sensor on the absorber plate in the collector and the other in a central location within the heated building. This configuration allows the system to turn on as soon as there is available heat within the collector and also shuts the system off when the building starts to get overheated. This is especially efficient when heating a building that is kept cool during the evening. In this case the collector can come on earlier in the day, plus the temperature differential

between the building and the collector is higher, thus increasing collector efficiency. A high-limit thermostat and relay can also be added for more precise overheating protection. With this option, when the thermostat reaches high limit, the power is cut off to the blower. Note that it does not hurt an air type collector to turn it off and allow it to stagnate.

Another way of regulating the fan in air systems involves turning the collector on at a preset temperature using some type of set-point thermostat. This type of control is the least efficient of any discussed here.

Another very important component in an air-type heating system is a set of dampers. We learned earlier that water thermosiphons within a tank. Well, air can thermosiphon within an air system and even within an air duct. This phenomenon is most apparent with roof-mounted collectors, but it also happens with wall-mounted collectors. At night, when the temperature drops, the air within the collector also gets cold. Because cold air is heavier than warm air, it can drop down the ductwork and flow into the building, and at the same time warm air from the building can rise up to the collectors and lose its heat. To prevent this, some type of damper must be installed on each primary duct going to and from the collector. Motorized dampers are used, and are set on the same electrical circuit as the blower. These dampers must have a positive seal or heat loss will occur. The only exception to the above damper rule is for wall-mounted PV-powered systems. In this case it may be possible to use a flapper damper that is placed on the lower cold air duct. These dampers can be made using hardware cloth and rip-stop nylon. This design also requires that the cold air inlet be rectangular and horizontal and the hot air duct be located at the top of the collector.

HEATING WATER WITH AIR COLLECTORS

Many homeowners and installers have attempted to heat water with air systems, especially during the 1980s. This system incorporated an air-to-liquid heat exchanger in the hot air duct, a water storage tank, and a circulating pump. The circulating pump came on whenever the blower came on. As hot air passed through the heat exchanger, the water theoretically was heated and the pump circulated the water from the storage tank through the exchanger. The overall efficiency of this system is very low. The air temperature is never very high, so the temperature differential between the air and water is small. This does not lead to very good heat transfer efficiency. Also, if you add up the electrical energy used to run the blower and the pump and the dampers, you don't come out very much ahead. But the biggest issue has been freezing heat exchangers. The heat exchanger must be located on the warm side of the damper. If the damper ever fails during freezing conditions, the heat exchanger will freeze and burst. Every single water heating system that I have seen that used air collectors has frozen at least once. In the worst case scenario, this is an especially ugly failure because when the exchanger thaws out, water shoots out of the rupture, and this water is hooked up to your main water supply. Say the exchanger freezes overnight and you don't know it. Everyone leaves the house for the day. The exchanger thaws out and water squirts out of the exchanger all day long. You come home to a flood and a big water bill. Heating water with air collectors is a bad idea.

ADDITIONAL COMMENTS ON AIR SYSTEMS

A few specialty applications of air systems worth mentioning involve solar chimneys. A solar chimney is a vertical duct that is glazed on the south side and is black on the inside. The duct is open on the bottom and the top. This duct is actually a solar air collector. When the sun shines on the collector/duct, the air inside is heated and rises. As the air rises and leaves through the open top it pulls fresh air into the bottom of the duct. Air circulation is created. When these collectors have been used for cooling, especially in warm climates, they are incorporated into the design of the building. The bottom of the duct is open to the building. As air rises in the duct on a sunny day, air is pulled out of the building. The air can be taken off the ceiling, where the building is hottest. Cooler air can be drawn into the building by the negative pressure caused by the solar chimney.

A promising technology that I have hopes for is the use of air rising in a solar chimney to spin an electric generator. Massive solar chimneys have been proposed in desert areas that could potentially generate significant amounts of power. Some of these designs have the solar chimney going up the side of a large hill or mountain. For current developments in this area, just enter solar chimney in your Internet search engine.

Chapter 7

CHOOSING THE RIGHT SYSTEM FOR YOUR NEEDS

~~~~~~~~~~~~~~~~~~~~~~~~~~~~➤

AFTER WE HAVE PUT IN PLACE our energy conservation measures, we are ready to decide what type of solar water heater will be best for us. There are several types of solar water heaters to choose from. Each type is slightly different and each is appropriate for different situations. However, like I stated before, I only recommend the ICS, drainback, and antifreeze systems. These three systems should cover all possible climates.

When selecting your system I caution you to always plan for the *worst case scenario.* Situations have a way of constantly changing, and what might be OK now, may not be in a couple of years. Remember that a properly installed solar water heater usually lasts decades. Think ahead! While we are talking about worst case scenarios, let's learn something from the past. You might recall from our review of the history of solar energy that in California's first solar boom some people had bad experiences with their solar water heaters because they purchased inferior and untested products. In most cases, they were sold on low system cost. Also recall that during the Florida solar boom the same thing happened for the same reasons. We experienced a similar case during the tax credit days of the 1980s. Most systems were good, but there were, and always will be, people out there trying to make a quick buck selling inferior products. This always seems to be the case when something becomes fashionable.

What can you do to avoid this experience? First, use your common sense. If a deal is just too good to be true, then it probably isn't good or true (there are no free lunches out there). Study this book to learn what it takes to make a quality system. Talk to people with experience. Be careful. If someone is trying to sell you something and they say that you must make the decision today (usually right after a sales pitch), don't do it.

I have found that 90 percent of the problems encountered with solar water heaters come from either using the wrong (usually inferior) component, poor workmanship, or choosing a system that is not appropriate for your climate. Of all the solar water and space heating systems I have installed over the last 20 years, only about 10 percent of them have ever required a service call.

Before selecting a system, you may want to find out if the rebate programs in your area require a specific type of system. Most utility rebates here in Wisconsin, for example, require that you install an antifreeze system because of the climate. Many rebate programs also require an SRCC-certified system to qualify. If you are dealing with a company that is new to the business, it would be wise to specify a certified system. However, experienced installers may offer a high quality system that is not certified, but is a wise investment anyway. Be sure to carefully check the credentials of the people you are dealing with.

Finally, the system you select may be determined by how it will be installed. For instance, a drainback system requires that the collectors are mounted above the drainback tank. If your only option for a collector array is on the ground, then a drainback system will not work for you. The following sections will help you choose an appropriate location for your collectors and help you size the system according to your needs.

Siting a Solar Energy System

Sometimes it seems silly to state the obvious, but here goes: solar collectors must be in the full, direct sun if they are going to work properly. This may seem obvious, but you wouldn't believe how many times prospective customers have told me that there is lots of sun where they are thinking of placing the collectors, only to find the spot in considerable shade. So, what is the bottom line for choosing a location for our solar collectors? Read on.

First, let's review how the sun shines down on the Earth. The movements of the sun across the sky determine our days as well as our seasons. The path of the sun changes every day. Here in Wisconsin (and everywhere on the planet), on the first day of spring (March 21) the sun rises directly in the east, at noon it is directly to the south and is above the horizon at around a 60 degree angle, and it sets almost directly west. On that day the sun is up for 12 hours and down for 12 hours. On the longest day of the year, the first day of summer (June 21), the sun rises in the northeastern sky, is directly overhead at noon (90 degrees from horizontal) and sets in the northwest. The first day of fall (Sept 21) is exactly like the first day of spring. The shortest day of the year is the first day of winter (Dec. 21), and the sun rises in the southeastern sky, is directly to the south at noon and is about 23 degrees above horizontal (low in the sky), and then sets in

the southwest. The height of the sun, in degrees, will vary depending on your geographical location. The further north you go, the lower the sun will be during the winter months.

Generally, solar collectors should face within 30 degrees of south, be mounted at an angle to the sun that will maximize their performance, and be in the direct sun (no shading at all) from 9 a.m. until 3 p.m. It is between these hours that a fixed point will receive 80–90 percent of all the solar radiation it receives over the whole day. Some solar installers advocate for full sun between only 10 a.m. until 2 p.m. In some cases this will be suitable, but for optimal collection, you should try to have full sun between 9 a.m. and 3 p.m.

To get the maximum performance from your solar energy system, the collectors must face the sun at noon. Both early and later in the day, the energy coming from the sun must pass through more of the atmosphere than during the middle of the day. This atmosphere is full of dust, water vapor, and moving molecules. All this stuff in the air interrupts and weakens the energy flow from the sun to the Earth's surface. The less atmosphere the sun's energy has to pass through, the stronger it will be. It may seem elementary, but solar collectors need to be in the full sun for most of the day. They will not work if the direct sun must pass through trees or other vegetation, even if the leaves are gone. A tree without leaves will block up to 75 percent of the sun's energy.

Collectors should face as close to true south as possible, but a variation of up to 30 degrees is generally acceptable and will not significantly reduce panel performance. To find south, you first have to find north and then look the other way. However, compared to a compass reading of north, true north is not the same everywhere. To find true north at your particular location, drive a stake vertically into the ground and watch its shadow. When the shadow reaches its shortest length, it is pointing exactly north. Unless you have some time to sit

FIGURE 7.1

Isogonic map of the United States

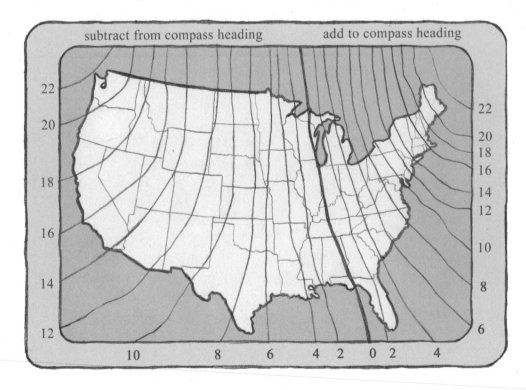

subtract from compass heading   add to compass heading

around staring at a stake in the ground, you probably want an easier way to find true south. You can also find the declination of the compass reading for your area from your local weather service, or by consulting an isogonic chart. An isogonic chart demonstrates how the compass reading of north will vary depending on global location. To use the isogonic map in Figure 7.1 you simply find your location on the map and note the nearest line. Those on the western half of the United States will have to subtract the given number of degrees from their compass reading and those in the eastern half will have to add them. For instance, if you live in southern California, you will need to subtract between 14–16 degrees from what your compass reads as south. Many compasses have a dial on them for you to make this adjustment.

FIGURE 7.2

The solar window from above

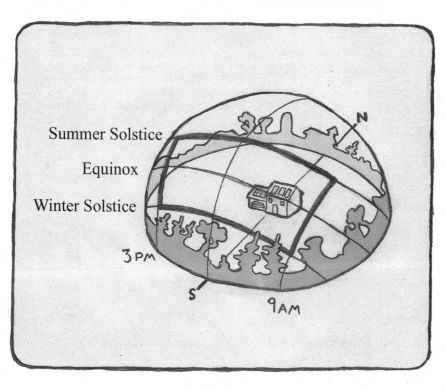

One way to determine if a proposed site is good is to imagine a big window in the sky directly to the south of your collector array. Imagine that all the sun's energy that will fall on your collectors must pass through this window. The solar window is defined by the path of the sun across the sky throughout the year. On June 21, the summer solstice, the sun takes its highest path of the year across the sky. The lowest path throughout the year is on the winter solstice, December 21. These paths of the sun define the top and the bottom of the window. The sides of the window are defined by where the sun is at 9 a.m. and 3 p.m. You want this solar window to be completely open and free of any obstructions.

Professionals may wish to purchase an instrument that can be set up at any location to show the solar window. There are two on the market today that work successfully. The first, the Solar Pathfinder, uses a plastic dome to reflect the solar window onto a sun chart. The sun path chart is selected based on your latitude and will have the appropriate sun angle for the different months of the year. You trace the shadows reflected on the dome onto the sun chart, giving you a concrete analysis of any obstructions and where they are. The Solar Pathfinder will also allow you to rotate the surface to account for magnetic declination, which is a handy feature. Recently re-released, the Solar Site

FIGURE 7.3

The solar window from ground level

Selector is your other option. It is composed of a semicircular base in which you attach a transparent sun path chart. This sun chart will vary based on your geographical location. To use the Solar Site Selector, point the instrument straight south and level and then look through an eyepiece located on the base. When you look through the eyepiece you will see the sun chart transposed over the actual solar window in the sky. The advantage of this instrument over the Solar Pathfinder is that you can actually see any potential obstruction instead of just their reflection. However, you do not end up with a concrete recording of the solar window.

If you are not willing to purchase one of these instruments, or you just want a quick way to determine obstructions, you just need to find out the angle of the sun during the winter solstice for your area. Figure 7.6 lists the approximate angles of the sun on December 21 at various latitudes at different times. Next, take a simple plastic protractor and a piece of straight wire. Bend the end of the wire and insert the bent end into the hole in the base of the protractor. Hold the protractor so that the flat edge is up allowing the wire to dangle down.

Benjamin Nusz

FIGURE 7.4
Solar Pathfinder

FIGURE 7.5
Using a Solar Site Selector

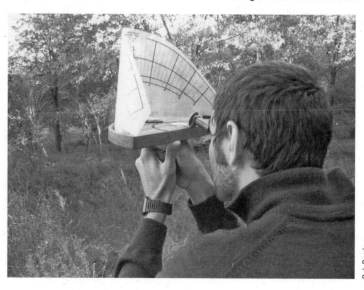

Bob Ramlow

Stand where you plan to locate your collectors and tilt the protractor so that the wire is pointing to the appropriate angle for your location for the appropriate time. Protractors typically have the 90° mark at the tip of the circle, so you will have to subtract your sun angle from 90° to get the "protractor angle." Next, point the protractor to where the sun will be during that time of day. At noon, the sun will be straight south. At 9 a.m. the sun will be just under 45° east and at 10 a.m. the sun will be just under 30° east. Look along the

straight edge of the protractor. If you see anything other than sky, you will have obstructions at that location during the winter solstice. Determining 2 p.m. or 3 p.m. should be done using the same angles. Now this is a rather crude method for determining obstructions, but it is simple enough that you can do it without much effort and without specialized tools.

If using any of the solar window assessment tools above reveals that more than five percent of the solar window has obstructions of any kind, you might want to re-evaluate the location. If the obstructions are tree branches, you may be able to remove

FIGURE 7.6

Sun angle chart during winter Solstice

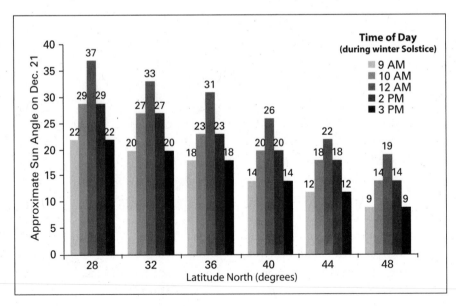

them. If the obstructions are only at the very bottom of the solar window, then the shadows will only be cast when the sun is lowest and the days are the shortest. This is the time of year when the solar energy system will produce the fewest Btus per day, so you may be willing to sacrifice some performance. This is a judgement call you will have to make. I always go for no shading. If the location where you would like to have collectors has just a few obstructions, it doesn't mean that solar is not for you. Sometimes you may be able to increase the size of your collector array to accommodate for shading. Personally, I always want the most out of every collector I install, so I tend to be kind of a stickler.

There are many locations that can be suitable for mounting collectors. They do not always need to be mounted on a south-facing roof or even on the roof at all. Figure 7.8 displays many of the collector mounting locations that have been successfully used in the past. You also have the option of mounting the collectors on a rack on the ground. However, these are not the only options: any place will work as long as it has a clear solar window and you adhere to the appropriate mounting methods.

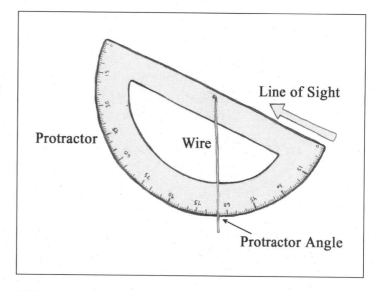

FIGURE 7.7
Protractor method

Roof-mounted collectors are more likely to be in a sunny location than ground-mounted collectors because they are higher in the air. As you evaluate your site for the best location for the collectors, you will see that the higher you go, the less shading you will encounter. Roof-mounted collectors also take up less room in your yard than ground-mounted collectors. Another advantage of roof mounting is that the length of piping required is often less than for ground mounts. Roof mounts are also less likely to be affected by vandalism than ground-mounted collectors. On the other hand, sometimes the dwelling is located in a shady spot and you would rather not cut trees around the house just for the collectors. In this case, you may be able to install a ground mount near the dwelling, in a better location to collect direct solar radiation. This is the case at my home. We have several very large maple trees in the yard directly to the south of our house. Those trees keep our yard and home nice and cool all summer and are also a haven for many birds. We ground-mounted our collectors away from the trees. A ground-mounted system often costs slightly more than a roof-mounted system because of the added cost of the rack which holds the collectors, as

FIGURE 7.8

Collector mounting locations

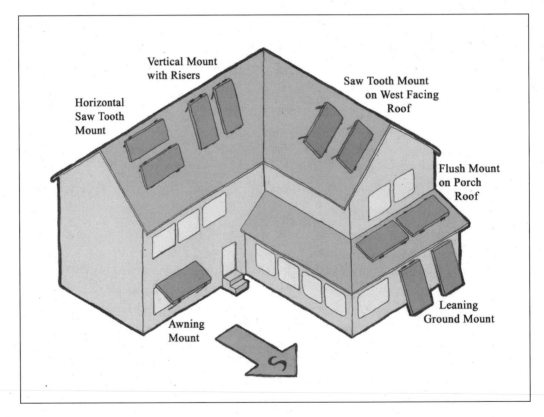

well as the added costs involved in trenching for the pipe runs. You will want to find the best location for your site. If there are multiple locations, all with clear solar windows, then you should consider other variables, such as aesthetics, cost, ease of installation, length of pipe runs, and/or weight load.

Sizing a Solar Water Heating System

Solar water heaters have been around for a long time, as we have seen in earlier chapters of this book. Over all those years of experience, some general sizing rules have emerged. I have found that using a simple sizing method in designing solar domestic water heating systems has worked very well. However, you will want to adjust the rules to meet your specific needs. There will also be times when the simple rules do not apply and you need to go through a series of calculations. This is particularly true in commercial applications. Often, these calculations will also need to be done when you are using evacuated tube collectors (because collector performance can vary widely between manufacturers). In the following, I will demonstrate both the standard rules of thumb and the method of specific calculations.

It should first be noted that in most cases you can't size the system to heat 100 percent of the water you use. You probably don't live in paradise, so you are going to have cloudy days. You will need to size the system to cover 100 percent of the load on a clear sunny day. If the system was sized to heat everything on a cloudy day, when you did have a sunny day you would be producing too much and overheat the system. Here in Wisconsin, a system sized to cover the entire load during the summer will typically cover 50 percent of the load in the winter and will typically heat about 75 percent of an annual load. Locations with milder winters can usually reach a higher annual percentage.

CALCULATING THE LOAD

To begin sizing your system you will want to start with the load and then work backwards. In order to determine how much water you need to heat, you need to know how much you use. On average, each person in a modern household uses 20 gallons of hot water per day. If you simply multiply the number of members in a household by 20 gallons, you will have your household's average daily hot water load. However, a number of factors will alter this estimate. As the number of members in a household increases, the average usage per person will decrease. This is due to economy of scale. For instance, you are more likely to wash full loads when you have many to do and people tend to take shorter showers if there is a line. In other words, a family of 5 typically uses 10 to 20 percent less hot water per person than a family of 2 or 3. Are you average or are you conservative in your hot water usage? If you have been successful in reducing your hot water consumption through conservation techniques, then you should reduce the estimate of your load. If you feel that you use more hot water than average, you will need to increase your estimate. For instance, if you have a spa tub that holds 40–50 gallons of water and you use the spa regularly, you have to take that into consideration. Temper the average with consideration for your water consumption habits.

Simply using the average is the quick and easy way of sizing. Like I said before, this is usually recommended as good enough for most cases. However, if you think that your load is much different than most and want to find out what it is specifically, or if you are designing a system for a commercial application, then you will have to put forth a little more effort. The most accurate method for determining how much hot water you use daily is to actually meter your present hot water consumption. For most residences, this would be an expensive test. To do this measurement, you would have to install a water meter on the hot water line of your home, run the test for a month, and then divide total consumption by the number of days in the test to get an accurate average daily hot water consumption figure. Most likely this is not an option. However, with a little clever investigation you can usually get a pretty close estimate of your average load without a water meter.

The best place to start is with your utility bills. If you use a natural gas water heater, look to see what your usage was for the month of July. Unless you have a pool heater or a gas range, you only used gas in the water heater, so you will know the number of therms you use on a monthly basis. A gas range will actually use less than a therm a month, so it usually isn't a big factor. In addition, we have a tendency to use less hot water in the summer than in the winter, so you will want to increase July's bill by around five to ten percent to get your average monthly consumption. You can't determine the average just by looking a monthly bill if you use propane to heat your water. You don't get a monthly bill. This would only work if you had your tank filled once the heating season was over and before the next heating season began. Then you could take the average over that time period.

If you have an electric water heater, you will not be able to determine your average consumption by your utility bills. You use too many other appliances to be able to isolate just the water heater. However, you can use a kWh meter. Usually, you simply plug the kWh meter into your outlet and then plug the appliance into the meter. Most electric water heaters are directly wired, so they will not have the plug necessary for most meters. If this is the case, you will want to ask an electrician for assistance. You don't always need to purchase the kWh meter, although they are rather handy

FIGURE 7.9

Average consumption chart

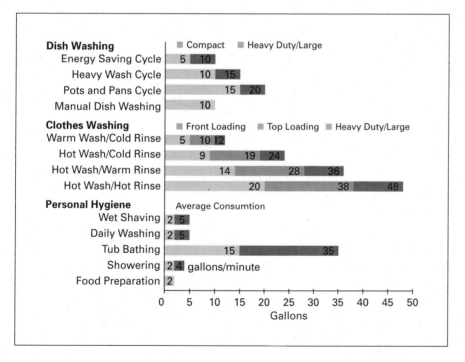

in reducing your electrical consumption. Many libraries have one that you can check out for a week or two and sometimes you can borrow one from your utility. I would suggest using the meter for an entire month, if possible, to find out an accurate daily average. Using it for only one day usually doesn't work because you may or may not wash the dishes or a load of laundry that day. These can be tracked as weekly tasks and averaged in to your daily load.

If the above methods don't fit your situation, you can still accurately estimate your average daily hot water consumption. To do this, you need to analyze the hot water consumption of each person in the household. Most people tend to underestimate the amount of hot water they use, so be careful to examine each person's daily habits carefully. To help you determine hot water consumption, the information in Figure 7.9 may be helpful. Using this method is usually very time consuming because you will need to take into account the flow rate of your faucets and the efficiency of all of your appliances. The numbers in the chart are just averages, so if you are looking for specifics you will have to do all of the hard work. To save time, I would suggest just using the average of 20 gallons per person described above.

SIZING THE SOLAR STORAGE TANK

Once you have a general impression of your hot water load, you can start to size the system components. You will want to start by sizing the solar storage tank. As a general rule, for every gallon of hot water you consume daily, you will want to have one gallon of solar storage. This rule is straightforward because the sun comes up daily and will be able to replenish what you use. An average family of three will probably want to have a 60 gallon storage tank to accommodate their 60 gallon/day load. Tanks come in a limited range of sizes, so you will have to select the one nearest to your need. I would recommend rounding up instead of down. It is okay to have a little bit too much storage, but if you have too little, you risk overheating the system.

SIZING THE COLLECTOR ARRAY

Knowing how many gallons you need to have in your storage tank will allow you to then size a collector array that will heat that amount of water. All you need to determine is the ratio between the number of square feet of collector to the number of gallons of solar storage. The ratio of collector area to storage capacity will vary depending on your geographical location because of the variance in the amount of solar radiation. Simply, the more sun a location receives, the more water you can heat with a square foot of collector. Figure 7.10 shows generalized solar zones of the United States. This will help you determine the amount of collector area you will need for each gallon of storage.

Gallons of storage per square foot of collector:

Zone 1: $1^1/2$ – 2 gallons
Zone 2: $1^1/4$ – $1^3/4$ gallons
Zone 3: 1 – $1^1/4$ gallons
Zone 4: $3/4$ – 1 gallon

Now you can appropriately size your collector array based on these ratios. For instance, using the ratio of 1 square foot of collector for every 1½ to 2 gallons of solar storage, a family of 4 in Arizona would want about 80 gallons of storage and 40–50 square feet of collector. Simply divide the storage capacity by the ratio number for your appropriate zone. If the same family lived in Wisconsin, they would want 70-80 square feet of collector. The zones are rather wide so there will still be a range within each one. Generally, the further north you go in a zone, the more collector area you will want per gallon. The following chart can be used

to summarize the simple sizing method. It assumes that the smallest storage tank you can find is 40 gallons, so the numbers for one person are the same as for two. Forty gallons is the smallest size you would want to install, anyway. You also need to remember that flat plate collectors come in only a few standard sizes. Most common are 4' x 8' and 4' x 10'. However, you can often find them in 4' x 6' or 3' x 8' or some other odd sizes. You will need to design the system using these limited sizes. For instance, if you need around 51 square feet of collector, you should use

FIGURE 7.10

Storage zone map

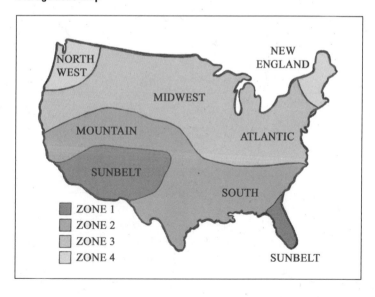

two 3' x 8' collectors to come up with 48 square feet. As long as you don't go too far over the upper limit of your zone, you should be okay.

Please remember that the simple sizing method uses averages. Also remember that a little more or less collector area will barely be noticeable because solar radiation is variable. If you are in doubt, always upgrade to the next larger size.

SIZING FOR EVACUATED TUBE OR CONCENTRATING COLLECTORS

Most evacuated tube and concentrating collector manufacturers will include sizing instructions with the manual. They will let you know the number of evacuated tubes or the size of the

concentrating collector you will need based on your load and your geographical location. However, if this is not provided, you can determine the size of the array if you know the efficiency of the collector. You will want to know the number of Btu per square foot per day that the collector will produce for your location. This can usually be found through SRCC. The average for most evacuated tube collectors is around 600–700 Btu/ft^2/day. Concentrating collectors have much more variance. Next you would want to convert your daily hot water load into daily Btu load. Instructions on how to make this conversion are covered more clearly in the section on calculating the space heating load. Take the total load and divide by Btu/ft^2/day of your collector, and you will get the number of square feet you will need. You can also use this method for flat plate collectors. Selecting the correct size of collector is much more vital for systems that use evacuated tube or concentrating collectors. They should be matched carefully to both the size of the storage tank and the load. With either type, too much collector or not enough load or storage can quickly lead to overheating the system.

FIGURE 7.11

Simple sizing chart

	Number of People in Household					
	1	2	3	4	5	6
	Size of Solar Storage Tank					
	40g	40g	60g	80g	90g	100g
	Square Feet of Collector					
Zone 1	20-26 ft^2	20-26	30-40	40-53	46-60	50-67
Zone 2	23-32	23-32	34-48	46-64	51-72	57-80
Zone 3	32-40	32-40	48-60	64-80	72-90	80-100
Zone 4	40-53	40-53	60-80	80-107	90-120	100-133

ADDITIONAL SIZING CONSIDERATIONS

The methods mentioned above base the size of the system on the size of your hot water load. However, sizing the system based on your current consumption may not always be the best idea. Consider the future: a solar water heater can last over 40 years. Do you plan to live in your home that long? You will need to consider the size of your home, how many bedrooms it has, and how many people could live there. If there are only two of you now, but you live in a four bedroom home, you may want to increase the size to accommodate a larger family, in case you sell it some day. However, be careful not to get carried away. A larger system is designed for a larger load, which keeps the system from overheating. Increasing the storage tank by 20–30 gallons is usually acceptable. You also need to consider the size of your own family. If you are planning on having children in the future and expect to live in the same home, you may want to size the system for additional occupants. Babies are messy; plan accordingly. It is important to mediate between your current situation and what you expect for the future.

You also need to take into account the clarity of your solar window and the orientation of your collectors. If your solar window has partial obstructions, but is the only option available, you will want to increase the size of your collector array by the percentage of your solar window that it obstructed. Similarly, if you are mounting at a location that is more than 30° away from due south, you will also want to increase the size of your array, probably in the range of one to two percent per degree that you are away from 30°.

Sizing a Solar Space Heating System

Let us start this section with a reality check. In most climates it is nearly impossible to heat a dwelling 100 percent with solar energy. The simple fact is that most climates have extended cloudy periods during the heating season. There are exceptions to this, like in some areas of the northern Rocky Mountains, but most people have to deal with clouds. Your particular climate will be the single biggest influence on sizing a system. The second biggest influence will be the heat loss of your building. The reality is that systems without storage can provide an average of up to 25 percent of the annual space heating load, systems with water storage can provide an average of up to 50 percent, and high-mass systems can provide an average of 75 percent or more, even in northern climates.

This section is about heating a building. Unfortunately, all buildings lose heat to the outside when the outside temperature is colder than the inside temperature. The way a building is constructed will determine the rate at which this heat is lost. Insulation retards the flow of heat, so the more insulation a building has the slower it will transfer the heat from inside to outside. Also, any holes in the building envelope will allow heat to escape the building and will also allow cold to easily enter. Little tiny cracks around windows and plates can add up, so the tighter the building was constructed, the less air infiltration will affect the heating load. We need to consider all the above factors when calculating the heating needs of a particular building.

A well insulated and tightly constructed building will require less energy to remain at a constant temperature than a poorly constructed and poorly insulated building. This brings us back to something we considered in the introduction: it is cheaper to conserve energy than to purchase it. Your first task is to reduce the building's heating load as much as possible. After you have done everything that is reasonably possible to make your building more energy efficient, then we can proceed with sizing a solar heating system. Often, with existing buildings, efficiency is upgraded at the same time as adding the solar heating system.

Sizing a solar space heating system is not as straightforward as sizing a simple solar water heater. There are many more variables to consider. As I have mentioned several times in the preceding chapters, sizing solar energy systems is not a precise science because the amount of solar energy striking any given square foot of the Earth is not constant. It varies from minute to minute as well as hour to hour and year to year. The size of the system will also depend on its configuration. Dump systems will require a smaller collector array than a system with storage. In addition, no two buildings are the same, so each will have its own heating characteristics. It is much easier to incorporate solar heating systems into new buildings than existing structures, where the options might be limited by how the building was constructed. I will cover these variables in the following sections and demonstrate how they will affect the size of the system.

The first step is to determine how much space heating we want. Do you want to go for the gusto, the whole banana, or would you just like to supplement with solar energy? In almost every situation, money enters the equation as a prime factor; obviously large systems require a larger investment than small ones. In addition, some system configurations will be limited in the percentage heat that can be delivered. Once we know our goal, we can proceed with determining the total heat load of the building for the whole year.

CALCULATING THE LOAD

Whether you are installing a system with storage or a dump system without storage, the first step will be to calculate the heating load for the building. If you have an existing house that has been heated with fossil fuels, then you should have a history of the amount of fuel it takes to heat it. Your bill will show the amount of energy you used for a certain period of time.

It is impractical to size a solar heating system to heat your building on the coldest day of the year, because

then for the rest of the winter your system would be oversized. A compromise must be made so your system is both cost-effective as well as having the ability to provide a significant amount of solar heat on an annual basis. Compromise means finding a middle ground. The middle ground of the heating demand of our building is the average daily heat load of the building over the heating season. Here in central Wisconsin the average building is heated on a daily basis from October 1 through April 30, so to get our average heating load we add up the heat used for this time period and divide by the number of days over the same period. For your location, start with the month when space heat is needed on a daily basis and end when that daily heating is no longer needed. We will define this time period as our official heating season.

Let's say that you presently heat with natural gas. You get a bill each month. Let's assume there were 30 days in that billing period. Natural gas is sold by either cubic feet or therms. Let's say your bill tells you that you used 100 therms of natural gas. Each therm contains approximately 100,000 Btu. To find the average heat load of the building for this time period, take the total amount of Btu consumed and divide by 30 days. This will give you the average amount of fuel used per day.

The important figure we need for solar heating system sizing is the average heat load over the official heating season, so add all the therms used during your official heating season and divide that total by the number of days in that time period. You will probably find that the average daily heating load for the official heating season is around 33 percent smaller than the average daily heating load for the coldest month of the heating season.

If you are building a new building or if you heat with electricity, you won't have the luxury of a track record to use to estimate your average daily heating load. It is usually difficult to track the amount of electricity you use for heating, even if you have a kWh meter. Instead, you will have to perform a heat load analysis of the building to get that figure. An architect usually does this calculation but many heating and solar contractors can do it as well. Heat load calculations provided by architects and energy professionals will be expressed as Btu/heating degree-day/square foot.

Most heating bills have another helpful bit of information: they display the number of heating degree-days during each billing period. A heating degree-day is a comparison of the outdoor temperature to 65°F. For example, if the average temperature outside for a 24 hour period was 60°F, then there would have been 5 heating degree-days during that day. This information can be used to determine how much energy is being used to heat the building. The bill will have the total heating degree-days listed for the billing period. For instance, if it was just over 30°F for every day in the month of December, you would have around 1,000 heating degree-days. You just need to divide the total amount of fuel used by the total number of heating degree-days. This will give you the amount of fuel used per degree-day. Convert that to Btu. When you have a heat load analysis done of your home the calculation will result in the number of Btu per square foot per heating degree-day. Now, that number alone is not sufficient to calculate your load. You need to multiply that figure times the number of square feet in your home to find out your total Btu per degree-day. You will next need to find out the number of

Comparative Thermal Values of Fuels

Natural gas:
100,000 Btu/therm.

Coal (good stuff):
12,000 Btu/lb.

Propane (LP):
91,000 Btu/gal.

Fuel oil # 2:
139,000 Btu/gal.

Electricity:
3,412 Btu/kWh

Gasoline:
125,000 Btu/gal.

heating degree-days for your geographical location. You can usually find them from your local weather service or sometimes from your utility. You will then want to tally the total for the months of your official heating season and then divide by the number of days in that season to get the average number of heating degree-days per day. Multiply the result times the Btu per degree-day for your entire home to get the average number of Btu per day you will need to heat your home. Presto! That's your load for solar sizing purposes.

SIZING THE COLLECTOR ARRAY

You will next want to match the size of your collector array to the size of your average load as calculated above. Because your load is based on the average number of Btu it takes to heat your home, you will need to find out how many Btu your collectors will produce. Through SRCC or on the panel's label you can find the average amount of Btu that a panel will collect. The amount collected will depend on the outside temperature and the amount of solar radiation. Typically, flat plate panels will collect around 700–1000 $Btu/ft^2/day$ on a clear day. In northern climates when

space heating is needed, the production is usually between 700–800 $Btu/ft^2/day$ during a cold clear day in the heating season. Use this estimate, or find out the specific amount for your collector to determine the total number of Btu/day for your collector array. For instance, a 32 square foot panel will collect around 24,000 Btu/day.

OK! Now we have two essential figures: 1) how many Btu we need for an average day during the official heating season, and 2) how many Btu per square feet a flat plate collector produces each day. Divide the total number of Btu needed per average day by the calculated Btu output per day of your solar collector in your climate, and you have the number of collectors needed to heat your building on the average winter day.

Here is a sample sizing calculation. The sample home is in Wisconsin and has an average official heating season therm usage of 636 therms for the months of October through April.

This size collector array should easily provide 50 percent of the yearly heating load for the average year in central Wisconsin for the sample home. As was mentioned above, for high-mass heating systems the solar contribution

Sample Sizing Calculation

Natural gas consumption for our official heating season: 636 therms

Days in heating season: 212 days

Average daily consumption: 636 therms /212 days– 3 therms/day

Convert to Btu: 3 x 100,000–300,000 Btu/day

Btu per square foot of collector: 750 Btu/ft^2

Divide total load by collector performance: 300,000/750–267 ft^2 of collector will be 25 percent higher. Many locations will see greater savings than the averages for Wisconsin.

Now you have a reference point to use to start deciding what size system you want, and what is practical. Undersized systems save you money. They do what they can and every little bit helps. Large systems are great because they can significantly impact yearly heating costs. We also need to ask: What is the point of no return? It doesn't make any sense to design for the worst day, because then all the rest of the days the system will be oversized and perhaps overheating. Another consideration we talk about soon is storage size and how that plays into our decision-making process.

SIZING SYSTEMS WITH STORAGE

Sizing a solar heating system with storage is more complicated than any sizing we have talked about so far. The variables are collector array size, storage size, expansion tank size, pipe size, and heat exchanger size. The biggest variable and the first question you have to ask is, how much of the annual heating load do I want the solar heating system to provide? You can use "rules of thumb" which have been determined over many years of experience,

or perform the calculations. I would suggest using both.

The discussion in the previous section concluded that for the sample house, we would need around 267 square feet of collector to provide enough Btu to heat the house for an average day. A solar storage tank sized to 1.2 gallons of storage for each square foot of collector will store enough heat for 24 hours of heating on the average day. Therefore an optimum sized collector array would be 267 square feet with a storage tank size of 320 gallons (267 x 1.2).

RULES OF THUMB FOR SIZING SOLAR HEATING SYSTEMS

As mentioned above, rules of thumb have been developed over the years from practical experience. They are a middle of the road estimate. Your particular conditions and climate will influence these rules. The hard part of using rules of thumb is estimating the heat loss of the building that will be solar heated. These rules are designed for the average home in a moderately cold climate like Wisconsin. In central Wisconsin we experience around 8,500 heating degree-days per year. You can get the average heating degree-days for your location and compare that to the central Wisconsin data. Also, if you have a highly insulated building, you must decrease the collector area. Likewise, if you live in a warmer climate, you should downsize the collector area. If you have a poorly insulated building, you must bring that building at least up to minimum standards of insulation. The big issue is overheating.

Most people's inclination is to get as large a system as is practical. They want to save the most money or have the greatest environmental impact they can. But some people back off from the biggest practical unit and decide on a smaller unit that will have some impact. It is helpful with this exercise to start with the biggest system possible and work backwards from there, if necessary. For sizing dump systems with no storage, we want to have one square foot of collector for every ten square feet of floor space on the main floor. We add 10 percent to the collector size for a two-story building. That sizing method should give us an annual solar contribution of 20 to 25 percent. When we add storage to a solar heating system, we can add more collector area to the array because we can store the excess heat for use later. Therefore, if we had

two square feet of collector for every 10 square feet of floor area on the main floor, we should expect up to 50 percent of our annual heating load to be provided by solar if we have adequate storage built into the heating system.

In most cases, the two-to-ten ratio will be the maximum collector array size you should consider. In a perfect situation it would be sunny every day. If this were the case, we could realistically get 100 percent of our heat from solar energy. But the reality is that we have cloudy weather. As I have mentioned several times, here in central Wisconsin we can experience over five cloudy days in a row during the heating season, and we can experience as few as five sunny days in the whole month of December. No matter how many collectors we have, they do us no good when it is that cloudy. For this reason, the maximum solar contribution we can expect for a solar heating system is 50 percent unless we are using a high-mass storage configuration. Where the solar heating system is being added to an existing building, I suggest sizing the system for a 50 percent annual solar contribution. Although this is the largest size, it is a realistic goal, and will provide a reasonable return on investment.

Using the above rules, you can get in the ballpark of the collector array size. Next we need to size the storage. This part is easy. We want between one and two gallons of storage for every square foot of collector area. I suggest having at least 1.2 gallons of storage for each square foot of collector. If you live in a moderate climate that experiences plenty of sunny days during the heating season, I suggest more storage.

SIZING WITH COMPUTER ANALYSIS SOFTWARE

Computer programs for sizing solar heating systems with storage are available. These are very convenient because they know all the right questions to ask, plus they have weather data for most locations on earth. All the formulas needed are written into the programs. All you have to do is respond to the questions the program asks, and out comes the answer. Because these programs are fast, it is easy to vary the parameters and see how changing storage size or array size will impact the estimated yearly savings. Unfortunately, these programs are costly and are not widely distributed.

When I first started in the renewable energy business over 25 years ago, computer programs were available that were simpler versions of the ones on the market today. Nonetheless, they were an invaluable tool for an inexperienced installer to use to properly size systems. After using these programs for a while, certain patterns emerged. This is where the rules of thumb came from, as well as from watching systems over the long term. For anyone considering going into renewable energy as a profession, I suggest getting a sizing program. For the individual who is interested in putting in one or two systems, I suggest trusting the rules of thumb to size your solar energy system.

The big thing to remember when talking about sizing a solar energy system is that we are at the mercy of the weather. The sun is a variable resource. Long-term averages are a reliable indicator of system performance, but no two years are alike. Now that our weather is being influenced by the greenhouse effect, even long-term averages become less reliable.

SIZING COMBINATION SYSTEMS

When sizing a combination domestic hot water and space heating system you must remember that when you get larger you can take advantage of economies of scale. A system experiences a certain amount of losses through the piping and related components. As the system increases in size, the losses decrease in relation to square feet of collector. For example, if your calculations say that you should have 8.5 collectors to heat the house, and you should have 2 collectors for your domestic hot water, that equals 10.5 collectors. Because the sizing you used for the domestic water portion assumed that would be a stand-alone system, you can now downsize that portion of the system by up to 25 percent because of the economies of scale. You would round down in this case to ten collectors.

Sizing Pumps

Several different types and sizes of pumps are used in solar water and space heating systems, but all of them fall into the general category of centrifugal pumps. In centrifugal pumps, veins or paddles are spun within an enclosed space. Centrifugal force and the action of the paddles on the liquid create the pressure. The units of measurement used in circulating pump sizing are head and flow. Head is the pressure produced by a vertical column of fluid plus the pressure produced by the friction of a fluid as it flows through a pipe and is expressed in feet of head. Flow is the volume of liquid per minute moving through the circuit.

The kind and size of pump will be determined by the needs of the particular system you are using. The biggest variety of pump sizes and types are found in the primary solar loop pump, or main circulator. The two main types of pumps used as the main circulator are high-head pumps and circulator pumps. Circulator pumps are designed to circulate a fluid within a closed loop. High-head pumps can pump fluid up an empty pipe a certain distance.

All systems that are full of solar fluid at the beginning of a heating cycle only require a circulating pump to operate properly. This is the case whether the closed loop is horizontal or vertical, and to virtually any height. For example, in a closed-loop antifreeze system, the solar loop is full of solar fluid at all times, so when the sun comes up in the morning, the pump only has to circulate the fluid throughout the system. The head in this type of system is only created by the friction of the fluid as it circulates throughout the solar loop piping and fittings. Circulating pumps are not designed to pump fluid up an empty pipe.

All systems that are empty of fluid in the solar loop when the system turns on require a high-head pump for the main circulator. For instance, in a drainback system the solar fluid is stored in the drainback tank when the system is off. When the system turns on, the main circulator must first fill the system with fluid and then continue to circulate the fluid throughout the system till it turns off, at which time the fluid then drains back into the reservoir. When filling a vertical pipe with fluid, the pump has to overcome gravity. This takes a pump with specifically designed paddles that usually operate at a higher speed. They also require more power than a plain circulating pump. The head in this type of system is created by both the vertical pressure of the fluid in the pipe and the friction caused by the movement of the fluid within the pipes. Be aware that centrifugal high-head pumps can only pump fluid to a height of 30 feet above the pump center in an empty pipe.

When sizing a pump you have to first determine whether a high-head or circulating pump is required for the type of system you are planning. Then you have to calculate the head and flow requirements of the system to determine the size of the respective type of pump you will be using.

For both types of pumps, you will have to calculate the flow requirement. This is determined by adding together the flow requirement of all the individual collectors. The flow rate used for testing solar collectors by the SRCC is 0.88 gallons per minute for each collector, so that is a good flow rate to design for.

The friction head pressure will also have to be calculated for both pump types. Friction head pressure is calculated by adding the standard friction head for each foot of pipe in the circuit plus the friction head pressure for each fitting in the circuit plus the friction head pressure for all the other components in the circuit. Friction head pressure for pipe and fittings can be found in plumbing design manuals and plumbing reference guides. The friction head pressure is determined by the size of the pipe or fitting and the flow rate. Other plumbing components are individually rated for head pressure and this information is often supplied with the specifications of the component.

Head pressure created by the pressure of the vertical height of the circuit will also have to be calculated, but only for systems that require a high-head pump.

Once you have calculated the total head pressure and flow rate of the circuit, you can then find the perfect pump for your job. All pumps have a pump curve, which is a graph that shows how the pump will perform at various flow rates and head pressures. These pump curves are supplied by pump manufacturers. You can find them on the Internet. Just type the brand and model of a pump you are considering into your search engine and you will find it. You can also probably jump from there to a manufacturer's main page and find other pumps they offer and see their pump curves as well. Wholesale suppliers also provide pump curves.

The pump curves that are provided by manufacturers represent the pumps' performance when they are pumping water. Propylene glycol is slippery and is harder to pump than water. For closed-loop glycol systems, you should calculate the pump size as if it were pumping water and then jump up one pump size to compensate for the glycol. This relates to your glycol dilution percentage we talked about in the components section. One reason to have as weak a dilution of glycol as possible is because the circuit will pump more easily the more water it has in the mixture.

SIZING AIR HEATING SYSTEMS

The important consideration when sizing air systems is to not oversize the collector. A balance must be met between collector output, building heat loss, and amount of thermal mass within the structure. In cold climates, the rule of thumb for sizing air systems is that ten percent of the footprint of the building equals the collector size. This is for an average building. For buildings with high thermal performance, the collector size should be reduced appropriately. For buildings with large amounts of thermal mass or poor thermal performance, the collector size can be increased accordingly. As a double check, it is helpful to compare the heat loss of the building to the rated collector output. Remember that the collector will be operating around four hours per day during the winter, so it has to do all its work during that relatively short time frame. In buildings with large amounts of thermal mass, collector output can exceed building heat loss by up to 100 percent. I defer to the collector manufacturer for any help they can give regarding sizing, especially when using transpired collectors. However, you will be doing similar calculations to those above for liquid-type systems by calculating the average heating load and collector output and matching the two.

Chapter 8

SYSTEM INSTALLATION

~~~~~~~~~~~~~~~~~~~~~~~~~~~~~~~~~➤

I N THE PREVIOUS CHAPTERS, we have described most of the technologies used today for solar water heating. The rest of the book covers details about installing and operating solar water and space heating systems. There are many principles shared by most of the systems we have talked about. As you have seen, most of them are variations on a similar theme. They therefore share many components and installation requirements. I will detail information about specific technologies as I work through the basic installation principles.

Installing Solar Water Heating Systems

A characteristic of every successful installation is proper and thorough planning. An installer with solid understanding of how a solar water heating system operates and what basic components are used for a particular system can visualize how that system will be incorporated into a structure. There are almost always several optional configurations to be considered. The options must be weighed against each other to determine the best of all possible locations for each component. The load analysis, which is the very first task in the process, will determine how many collectors will be used. The solar window will determine the best location for the collectors. Both the location of the collectors and the layout and type of building will direct where the main solar loop piping will be located. The area where the existing water heater is located will most often identify the area where the heat exchanger and the solar storage tank will be located. You need to visualize the system with all its parts already installed on the building.

Once it is decided where all the components will be located, it is time to make a list of all the necessary parts. A good method of making sure you have them all is to start at the top and work your way down. It is also very helpful to make a sketch of the system, which will help identify the needed components. You have already determined how many collectors will be used. Next, what type of mounting hardware will be used for the collectors? For roof-mounted collectors, determine how the piping will penetrate the roof, and identify the proper roof flashing. Measure the pipe runs next, from the collectors all the way to the heat exchanger and storage tank, and be sure to add the appropriate pipe insulation. Now go back and

identify all the pipe fittings that will be needed, including elbows, tees, unions, adapters, and pipe hangers. While you are doing this, visualize how you might actually get the pipe installed in the areas you have identified. Is there enough room to maneuver the pipes into their final location, or will you have to cut short lengths to accommodate the installation? Make sure you have plenty of couplers to join straight lengths of pipe together. The size of the collector array will determine what size storage tank will be required and what size heat exchanger or heat exchange module you will need. If you are not purchasing a pre-plumbed heat exchanger module, you will need to identify all the components required to build your own. These parts may include boiler drains, check valves, drainback tank, thermometers, pressure gauge, pressure relief valve, and others.

Once you have all the piping figured out, you can then calculate how much solar fluid will be required. Please refer to the system schematics in earlier chapters to make sure you have everything you will need. Also remember never to mix piping of dissimilar metals when installing a solar water or space heating system — or any plumbing system, for that matter. Compatible metals include: copper, stainless steel, brass, and bronze. Non-compatible metals are cast iron, galvanized steel, black iron, and aluminum. (Actually, all iron and steel pipes are compatible among themselves, but are not compatible with the other types of metals mentioned.) Whenever non-compatible metals must connect together, as where the copper piping attaches to a steel storage tank, always use a di-electric union at each connection point.

Professional installers and plumbers usually carry a large assortment of parts with them to every job site. I used to call my installation truck my rolling hardware store. Even if you have your own rolling hardware store, it is very important to check your inventory before you leave for the work site to make sure you have everything. Having to run to the plumbing supply house several times will add frustration and additional costs to any installation. I recommend having extra parts on the work site, especially pipe fittings, soldering supplies, solar fluid, and consumables like fasteners.

Some states or municipalities require professional installers to be licensed.

FIGURE 8.1

Pipe wrench

FIGURE 8.2

Channel locking pliers

Tools

- soldering torch with extra fuel
- pipe cutter
- de-burring tool
- fitting brushes
- soldering paste brush(es)
- charging pump and hoses
- pressure gauge (automotive type)
- tape measure
- marking pens and pencils
- hacksaw
- assorted screwdrivers
- assorted pliers
- electric drill(s), including battery-operated type
- assorted drill bits, including assorted hole saws
- hammer(s)
- assorted pipe wrenches
- channel locking pliers
- assorted crescent wrenches
- assorted socket wrenches
- reciprocating saw with assorted blades
- putty knife
- utility knife with extra blades
- level
- angle finder
- square
- wire stripper
- tin snips
- crowbar
- step-stool or short folding step ladder

- six to seven foot folding stepladder
- extension ladder(s)
- roof jacks with planks
- rope
- first aid kit
- five gallon pails
- work gloves
- safety glasses
- flashlight and assorted work lights
- extension cords
- sawhorses
- camera
- crane or lift (optional)

Supplies

- lead-free solder
- soldering paste
- grit cloth
- pipe dope
- Teflon tape
- assorted fasteners:
  - screws
  - nails (including roofing nails)
  - stainless nuts, bolts, washers
- roof cement
- caulk
- assorted zip ties
- electrical tape
- insulation adhesive
- identification labels
- rags
- sensor wire or electrical wire
- wire nuts or electrical solder

FIGURE 8.3

Adjustable crescent wrench

FIGURE 8.4

Torpedo level

Be sure to check with local authorities to find out if any licenses are required. Building permits may also be required, so check with the local building inspector about obtaining a permit.

The last part of the planning process is to make sure you have all the right tools and supplies to facilitate the installation. The following is a list of typical installation tools and supplies. A good solar installer will have all of them handy in his or her toolbox.

Handling Solar Collectors

SHIPPING

As I said before, most installations will start with the collectors. However, first you need to get them. Collectors are typically shipped stacked on a wooden pallet. When you receive your shipment of collectors, carefully inspect each collector for shipping damage. An obvious indication of damage will be broken glass. Other damage could be dents to the collector frame or back. Any damage should be written on the bill of lading. If the damage is severe, you may choose not to accept the shipment. If the damage is minor, you can accept the shipment and negotiate a settlement with the shipper for repair costs. As long as the collectors are structurally sound, they will probably be OK.

You usually have two options for receiving the shipment. You can have the collectors shipped directly to your site or you can pick them up at the local trucking terminal. If you have them delivered, you will have to unload them off the back of a semi-trailer. This is easy if you have a loading dock and even easier if you have a forklift. If you don't have access to a loading dock, the next best thing is to back a truck up to the tailgate of the semi-trailer. If you are receiving more than one or two collectors, it will probably be easiest to open the crate while it is still on the semi and then unload the collectors individually. Most semi-drivers will help you unload, but I suggest having two people available when delivery is expected. Also have a tin snips and crowbar ready to open the crate. Remember that the drivers have schedules to keep and usually don't have all day to fool around unloading one shipment. When you order your collectors, you can request a phone call from the trucking company alerting you to the exact delivery time. There may be a small fee for this service, but it is usually worth it because you can be totally prepared to accept your delivery and have your help there.

The other option for receiving your collectors is to have them kept at the local trucking company terminal in your area until you can pick them up. There are several advantages to this option. First, shipping is cheaper if the trucking company does not have to deliver right to your place. Second, you don't have to wait around for your delivery to arrive. You can go to the terminal at your convenience and take your time uncrating the collectors. You will have to have access to a truck or trailer that is large enough to haul the collectors to your site if you choose this option. Note that most trucking firms will start to charge you a storage fee if you do not pick up the collectors within a reasonable time, usually one week.

Each collector weights 85 to 140 pounds. If they will be transported a short distance, they can be laid flat, with the glass side up. If they need to be stacked, place cardboard between the collectors, taking special care not to let them slide around on each other. Securely tie the collectors down for transportation in such a way that they will not shift. They can be easily damaged so they must be securely tied down before transporting them.

On one hand, collectors are very tough pieces of equipment which will last decades on the roof of your dwelling. On the other hand, they are easily damaged during handling and transportation. The collector manifolds and glazing can be easily damaged and the glazing strips can be easily scratched. To prevent damage, always handle the collectors only by the frame, never by the manifold pipes protruding from the frame. If you pick up the collectors by these manifolds, they can bend and get out of round, causing problems later in the installation process. If the collectors have been sitting in the sun, they can get very hot, especially the manifolds. Always use gloves.

The best place to store collectors prior to mounting is inside a building and out of the weather. Collectors can be stored outside if well covered. Never leave the collectors exposed to the direct sun for any period of time because they can (and will) get very hot and can cause burns if touched with bare hands. Don't cover them with clear plastic because they will get too hot.

Most collectors are shipped with plastic caps on the manifold ends.

These are intended to keep moisture and dirt from entering the collector. If the collector is exposed to prolonged sunlight, these manifolds can get very hot, as mentioned above. The plastic caps can either blow off or melt onto the manifolds, causing additional problems during final installation — all the more reason to keep the collectors well covered. Some manufacturers also tape a plastic bag to the glass, which contains installation instructions and warranty cards. Remove these as soon as possible so they won't get lost or melted to the collector.

GETTING THE PANELS ON THE ROOF

Carefully plan how you are going to get the collectors up on the roof. Always remember, safety first! Don't try lifting the collectors when it is very windy or gusty. Always keep bystanders, especially children, away from the action. Whenever possible, use scaffolding, roof jacks, and planks when working on a roof. Always use safety ropes. I have found that wearing sneakers or running shoes gives a better grip than leather soled shoes. Rubbers also give a good grip. Heavy lugged boots tend to tear up shingles, especially when the roof is hot. Avoid walking repeatedly on the same area of the roof to keep wear to a minimum. The use of planks or plywood on top of the shingles also saves roofing from wear. Don't walk on the ridge. The steeper the roof, the higher the danger. Make sure your safety ropes are securely tied to something that will hold when necessary. Take your time and don't rush.

The easiest and safest way to get collectors up on the roof is to use some mechanical lift such as a bucket truck, crane, forklift, or roof hoist. Many roofers today have devices to place shingles on roofs and these can work well.

Manually lifting collectors up to the roof is also possible. I suggest having at least three people for this plan. Set two ladders up next to each other and leading to the roof. Have one person on the roof and two on the ground. Install the mounting brackets onto each collector. Lean a collector on the ladders with one end on the ground. Have a rope long enough to reach from the roof to the ground and tie the lower end to the collector. Tie the rope to the mounting brackets, never to the manifold pipes. When ready, the workers on the ground should slide the collector up the ladders while the person on the roof pulls

on the rope. As the collector slides up the ladders, the two workers can walk it up till the collector reaches the top of the ladders. At this point the person

FIGURE 8.5

Manually lifting collectors

Leif Ramlow

on the roof needs to set the rope aside and grab the top of the collector. The collector then can continue to move up until half of it is above the top of the ladders. Now pivot the collector top down towards the roof. You must be careful at this point to not rest the back of the collector on the top of the ladders, as this will dent it. Hold the collector just above the top of the ladders. The worker on the roof can move up the roof with the top of the collector till the bottom of the collector passes the ladders. The two workers on the ladders must hold the collector securely throughout this whole process. After the collector is lying on the roof, the workers on the ladders can climb onto the roof and help the other worker move the panel to its final location. If the roof is steep, have roof jacks and a plank set up near where the bottom of the collectors will be mounted. The roof jack and plank will keep the collector from sliding off the roof.

Mounting Solar Collectors

A previous section was devoted to siting your solar collectors. This process identified potential locations for your collector array. After considering all the variables you have typically chosen to mount the collector array on a roof or on a rack on the ground. We will address each of these mounting methods. Much of what follows describes the mounting of flat plate solar thermal collectors. As was mentioned before, the temperature operating range of flat plate collectors is best suited for most applications, so I will not discuss mounting evacuated tube collectors. However, most mounting tips will still be relevant. If you are mounting evacuated tube collectors, carefully follow the manufacturer's instructions.

COLLECTOR TILT

Before you begin to mount your collectors you will need to know their collector tilt — the angle the surface of the collector makes with the horizon. As we have mentioned many times already, solar collectors will perform at optimum efficiency and maximum output if directly facing the sun. We have observed how the sun tracks across the sky from morning to night. Wouldn't it be nice to have our collectors face east in the morning to collect the early morning sun, then follow the sun's path all day long till sunset? A good idea, but unfortunately our collectors need piping to bring the heat they collect into the house. There are no flexible piping products available today that would allow them to follow the sun. We therefore have to utilize a fixed array. Other solar technologies like photovoltaic arrays can track the sun's path across the sky because they deliver electrons through wires, which are flexible.

We have previously noted that the angle of the sun at noon will vary depending on the season of the year, being lower in the winter and higher in the summer. Because we cannot seasonally adjust the angle at which our collectors are mounted in order to maximize collector performance, we need to choose a permanent mounting angle that will be best for year-round performance. This mounting angle will be a compromise between summer and winter optimums.

Collector tilt will be determined by several factors. The chart in Figure 8.6 describes basic rules for determining collector tilt.

For a solar hot water system that will be used year-round, a tilt angle equal to the site's latitude is considered best. You can get your latitude from a map or from your local weather bureau office. For a solar water heater that will be used only during the summer, subtract ten degrees from your latitude to get your optimum tilt angle. Because solar heating systems are intended to work best during the winter months when the sun is low in the sky, the collectors must be tilted to maximize this low sun angle. Mount your collectors at an angle that equals your latitude plus 15 degrees. Here in Amherst, Wisconsin, our latitude is 45 degrees, so we mount our collectors at a 60-degree angle. People often comment that if their system is a combination space heating and water heating system this high angle might decrease its efficiency during the summer when they still need hot water. This is true, but there will be many collectors in this system, many more than are required to heat the domestic hot water. So the bottom line is that you will still have more hot water than you could possibly use with the collectors mounted at this tilt. If you plan on a year-round system for just domestic hot water and you live in an area that gets significant snow, your collectors should be mounted at an angle over 40 degrees to help the snow slide off the collectors.

**System Use
(Example @ 45 degrees latitude)**

Domestic hot water
Year-round usage
Title Angle: Latitude
45 degrees

Domestic hot water
Summer use only
Title Angle: Latitude-10 degrees
35 degrees

Space heating/
Domestic hot water
Combination
Title Angle: Latitude+15 degrees
60 degrees

FIGURE 8.6

Collector tilt chart

COLLECTOR MOUNTING

Flat plate collectors are a lot like big sails. Because of this, they need to be securely mounted. The mounting system needs to withstand winds of 100 miles per hour and gusts up to 150 miles per hour. In coastal areas where hurricane force winds are possible, extra precautions should be observed when mounting collectors. The wind pressure on a solar collector, especially when blowing from the north, will create pressure in every direction, including lift. This lifting force of the collector is very strong. Plan accordingly. I have seen people install collectors using screw-type fasteners and have their collectors blown off their roofs. It was not a pretty sight.

Collectors' mounting systems can be adapted to varied situations. All manufacturers of thermal collectors also make mounting hardware for their collectors. Mounting brackets, usually four per collector, are attached to the collector frame. These brackets can either bolt directly to a roof, be bolted to extension legs, or be bolted to a rack. Every manufacturer has a unique way of attaching a bracket to their particular collector, so you want to purchase the brackets for your collector from the collector manufacturer.

The brackets usually can be located anywhere along the collector frame, and can be tightened securely in many positions. Often, the bracket is shaped like an L. Brackets can be located on the top and bottom or on both sides of the collector. Usually, brackets bought from the manufacturer can be used for parallel roof mounting and also for rack mounting. Sometimes there are special brackets used for attachment to adjustable legs.

If your collectors do not have mounting hardware included, and the manufacturer is no longer in business, you will have to come up with your own method of attaching the collectors to the roof or rack. Some collectors have a lip or flange running around the perimeter of the collector and you can often drill and bolt directly to that flange. If no flange is available, you will have to attach homemade brackets to the sides of the collector. When attaching the bracket to the side of the collector, it is important not to drill a hole into the absorber plate or any piping within the collector. Look carefully through the glass to see where these are and drill accordingly to miss them. I use aluminum L-brackets in this situation. I suggest using stainless tech screws to attach the L-bracket to the collector. The sides of a collector are the best place to mount these L-brackets.

Flat plate collector frames and mounting hardware are typically made of aluminum. Because galvanized bolts and washers rust quickly when they contact aluminum, always use stainless steel for all fasteners. If stainless fasteners are not available, use a neoprene or rubber washer between all dissimilar metals.

Many flat plate collectors have weep holes on one end. Most collectors are rectangular, and the weep holes will be on one of the short ends. Mount the collectors with the weep holes on the bottom. These weep holes allow moisture to escape from the collector and allow the collector to maintain atmospheric pressure. Otherwise, they could build up pressure when they get hot, which could pose problems.

ROOF MOUNTING

When identifying the final location on the roof for the collectors, be sure to notice any potential shadows that could be cast by chimneys or other sections of the roof onto the collectors and adjust the array location accordingly. Also look for shading from trees or other buildings. It is usually best to try and have the top of the array as far up the roof as possible. This will allow easier fastening of the collectors to the roof structure and will also reduce the amount of snow that could accumulate behind the collectors in high snow areas.

It is very important to identify all power lines in the working area. It is very dangerous to be handling collectors near electrical lines. You risk electrocution if solar collectors come in contact with electrical lines. Avoid locating collectors where there is a possibility of contact. If the very best array location has power lines nearby, consider having the power lines moved before attempting to mount the collectors.

Make sure that the roofing materials are in good shape before mounting the collectors. If the shingles need to be replaced, it is wise to do this before mounting the collectors. Wood shake and tile roofs are very fragile, and extreme care must be taken when walking and mounting collectors on them. Remember to use planks and plywood to protect roofing. Most installers like to get as much of the roof work as possible done early in the day, as it is often cooler then. Also avoid walking on a hot roof because shingles are soft when they get hot and are more susceptible to damage then. It is typical to install a set of temporary roof jacks and a plank just below where the bottom of the collectors will be located. The roof jack and plank arrangement makes a stable place to stand as well as providing a safety barrier to keep collectors from sliding off the roof.

Flat plate collectors only weigh about three to four pounds per square foot, so they do not pose a significant weight problem for most roofs.

Nonetheless, it is always good to inspect the roof structure to make sure there are no rotten roof members. Someone will be walking around up there and you don't want anyone falling through the roof. ICS collectors, and some evacuated tube collectors, can pose a significant weight problem, because these collectors can hold a large amount of water right inside the collector. Careful calculations should be made to guarantee the roofing system will hold the additional weight. Additional roof bracing may be required. If you are not qualified to do these calculations, please consult an architect or builder for advice.

FIGURE 8.7

Hardware mounting methods

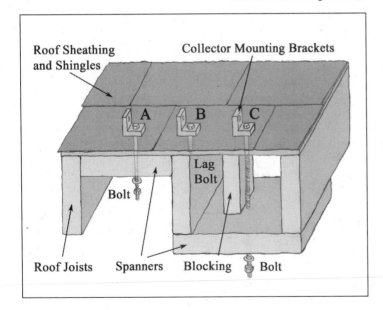

HARDWARE MOUNTING METHODS

As mentioned above, the collector array must be securely attached to the roof in order to withstand wind speeds of 100 miles per hour or more. The three most popular methods of attaching mounting hardware to roofs are: (A) using lag bolts or regular bolts secured through the roofing materials and into spanners installed under the sheathing and between the rafters, (B) with lag bolts secured through the roofing materials and into the rafters and (C) using long bolts secured through the roofing materials and through spanners located on the underside of the rafters.

Of the three options mentioned above, A is the strongest and is the method recommended for all installations if possible, especially if high winds are experienced at your location. Options B and C are also very strong and can be used in most locations and for attaching extension legs.

Option A shows a roof bracket that is bolted to a spanner located under the sheathing. The spanner should fit tightly between the rafters. The spanner can be a two-by-four or two-by-six. The spanner should be pushed tight to the sheathing and secured to the rafters with two nails or screws on each end of the spanner. Pre-drill your bolt holes slightly larger than the bolt size and use a large washer on the end of the bolt. This method works well if there is to be a finished ceiling attached to the rafters.

Option B shows a roof bracket that is lag-bolted into a rafter. Always pre-drill holes for the lag bolts. If lag bolts are screwed into rafters without pre-drilling, there is a good chance you will crack the rafter, resulting in a weak fastening. Use a drill bit that is smaller than the lag bolt so there will

be wood left for the threads to bite into, and drill the hole slightly shallower than the length of the bolt so the end of the bolt screws into solid wood. If you are using a spacer between the mounting bracket and the roof, drill a hole completely through the spacer using a drill size slightly larger than the largest diameter of the bolt so it will slide through the spacer easily.

Option C shows a roof bracket that is bolted to a spanner located on the underside of the rafters. A long bolt or threaded rod is used to fasten the bracket to the spanner. Wood

FIGURE 8.8

Determining roof pitch

Roof Pitch	Tilt Angle (degrees)
0:12	0
1:12	5
2:12	9
3:12	14
4:12	18
5:12	23
6:12	27
7:12	30
8:12	34
9:12	37
10:12	40
11:12	43
12:12	45
14:12	49
16:12	53
18:12	56
20:12	60

FIGURE 8.9

Converting roof pitch to degrees chart

spacers that are the same dimension as the rafters are used between the sheathing and the spanner to prevent the roof sheathing from caving in when the nut is tightened.

For flat roofs, a rack needs to be built to hold the collector array at the proper angle and orientation. The rack will be similar to ground-mount racks. For flat roofs, it is best to contact the installing roofer or a roofing professional. The roofer will install a curb or pitch pot, which are waterproof piers to which you can attach the roof brackets. Remember to consider the strength of the mounts, which have to hold the collectors down during high wind conditions. Remember, too, that the panels want to lift up during high winds, so plan accordingly.

Whenever a hole is put into the roof, adequate sealant must be used to assure a leak-proof installation. Use liberal amounts of silicone sealant or roofing cement inside the roof penetrations, under mounting brackets and shims, and around all fasteners.

ROOF PITCH

When mounting collectors on a roof, you must determine the pitch of your roof. Roof pitch describes the number of inches a roof drops per foot. You can determine your pitch by using a level and a tape measure. See Figure 8.8. Place one end of a level on the roof and hold it perpendicular to the roof and level horizontally. Measure 12 inches out from the roof on the level, and that will be your run. Now measure the distance vertically from the 12 inch mark on the level down to the roof, and that will be your rise. For instance, you might measure four inches down. This equates to a 4:12 pitch or approximately 18°.

Use the chart in Figure 8.9 to get your tilt angle from the roof pitch. Solar professionals may want to purchase an angle protractor, also called an Angle Finder, for a quick way to find the angle. To use an angle protractor, you simply place it on the roof and the needle will swing to the appropriate angle.

RISER LEGS

If the roof pitch equals the angle you have determined as optimal for your area, you can mount the collectors flush to the roof. In this case, the collectors will be held between two to three inches above the roof using mounting brackets and spacers (if needed). If the roof pitch is less than the optimum mounting angle, the collectors will need to be tilted. This is accomplished by raising the top of the collectors with riser legs that are attached to the roof with mounting brackets.

The length of the riser legs will vary with each installation according to the pitch of the roof, the collector mounting angle, and the length of the collector. Below is a chart you can use to determine the length of the riser legs for the collector. This particular chart is intended for 48-inch-wide collectors mounted horizontally. If you plan to mount a 4' x 8' collector vertically, you will need to double the length. For a horizontal 4' x 10', multiply the figure by 2.5. Negative numbers in the chart mean that you need riser legs in the front of the collector instead of the back. For odd-shaped collectors, you can use the following trigonometric calculation to determine the length:

$$S = 2L \sin \frac{T-P}{2}$$

Where:

S = Standoff Length

L = Vertical Length of Collector

T = Tilt Angle of Collector

P = Pitch Angle of Roof

FIGURE 8.10
Angle protractor

INSTALLING ROOF MOUNTS

Before attempting to hoist the collectors up to the roof, it is important to plan how you are going to attach the collectors to the roof, and to have all the necessary components assembled beforehand. You will need access to the underside of the roof during the collector mounting process, so have this access ready beforehand as well. Likewise, have all your tools gathered together before starting.

Once you have identified a potential roof area for collector location, it is time to lay out a plan on the roof to make sure that you have enough room for the array. Measure your collectors or look on the spec sheet for the collector to get the exact collector size. We talk about generic collector sizes like 4' x 8' or 4' x 10', but actual sizes vary for every model. Collectors are typically slightly larger than the generic size. Also calculate the additional distance that will be required between collectors. Most flat plate collectors have about 1.5 inches of manifold pipe protruding out their sides near the top and bottom. These pipes facilitate plumbing collectors together. Some collectors have factory-installed unions attached to these manifold pipes, which can add up to an additional inch of space required between collectors. In most cases, you will need three to four inches between collectors when mounting. You also need to have around 12 inches of roof space on each end of the collector array for pipe runs and roof penetrations.

FIGURE 8.11

Determining riser leg length

ROOF SLOPE		COLLECTOR TILT							
Pitch	Angle	30 degrees	35	40	45	50	55	60	65
Flat	0 degrees	25	29	33	37	41	44	48	52
1:12	5	21	25	29	33	37	41	44	48
2:12	9	17	22	26	30	34	38	41	45
3:12	14	13	17	22	26	30	34	38	41
4:12	18	10	14	18	22	26	30	34	38
5:12	23	6	10	14	18	22	26	30	34
6:12	27	3	7	11	15	19	23	27	31
7:12	30	0	4	8	13	17	21	25	29
8:12	34	-3	1	5	9	13	17	22	26
9:12	37	-6	-2	3	7	11	15	19	23
10:12	40	-8	-4	0	4	8	12	17	21
11:12	43	-11	-7	-3	2	6	10	14	18
12:12	45	-13	-8	-4	0	4	8	13	17
14:12	49	-16	-12	-8	-3	-1	5	9	13
16:12	53	-19	-15	-11	-7	-3	2	6	10
Vertical	90	-48	-44	-41	-37	-33	-29	-25	-21

Length in inches

Draw a diagram of your collector array on paper and calculate all measurements. Now go up on the roof with your marking crayon, chalk line, tape measure and square. Lay out where the collectors will be mounted. Make sure you are not above the overhang with any part of the array. All collectors should be pitched slightly towards the supply end to facilitate system draining. This is critically important in drainback or draindown systems, where the pitch should be exaggerated. For closed-loop antifreeze systems, the collectors should be pitched at 5° and for drainback systems, the collectors should be pitched at 15°. Mark the corners of the array and then snap a chalk line around its perimeter. When snapping the lines that are perpendicular to the roof rafters, pull the line tight and notice if the roof has developed sags. It is common for roofs sheathed with OSB or plywood to sag between rafters. If there is significant sagging, you will need to place shims between the roof and mounting brackets to ensure a proper installation, because collectors which are plumbed together must be kept in a straight line. If parallel collectors are not straight, getting them to seal together is difficult at best, and often impossible. Shims can be made of cedar or treated wood.

If there are multiple collectors in the array, start on either end and get the first collector mounted.

When starting the collector mounting process, first identify where the bottom mounting hardware will be located and attach the mounting hardware in the appropriate manner. If the collectors will be flush-mounted, proceed to install the upper mounting hardware at this time as well. If the collectors are to be tilted up at an angle greater than that of the roof, locate and install these mounting hardware feet as well. If you are inexperienced, I suggest that you mount one collector hardware one set at a time, and after that set is installed, proceed to mount the first collector. (Experienced installers often install all the mounting hardware sets before they attach any collectors.) When measuring for subsequent collectors in the array, don't forget to add for the distance between the collectors. Leave the mounting hardware fasteners loosely attached to the collectors until all collectors are mounted and then go back and tighten all the fasteners.

There are two methods of attaching collectors together. One uses brass unions; the other uses copper couplers. I prefer to use solid brass unions between collectors. The unions are soldered onto the collectors while they are still on the ground or in the shop. Some collectors come fitted with unions from the factory. The use of brass unions on the collectors minimizes the amount of soldering up on the roof, and multiple collector arrays are much easier to install when unions are used. Also, if you ever have to replace an absorber plate, it is much easier when brass unions were used during installation. If you are using unions, just snug them up as you are mounting the collectors and do the final tightening after all the collectors are in place. The best ones use an O-ring in the union.

Copper couplers are harder to use for several reasons. First, the copper collector nipples must be cleaned and fluxed before the collectors are slid together. Next, sliding the second and subsequent collectors together (necessary when couplers are used) is often difficult after the collector is already attached to the mounting hardware. If you attempt to slide the collectors together before fixing the collector to the mounting hardware, it is often very difficult to get the collector to line up with the mounting hardware. If you are using couplers, wait to solder them to the collectors until all the collectors are mounted and pushed tight together.

It is extremely rare to have the copper nipples line up perfectly when attaching collectors together in an array. Of course, they will be close, but they need to be perfectly lined up when making the connections. This is particularly important when using couplers, where they have to be exactly lined up. Brass unions are slightly more forgiving. Note that the copper nipples are relatively soft and bend quite easily. If you notice a deformed nipple when uncrating the collectors, attempt to get it straightened out before taking the collector up to the roof. If couplers are used, the nipple must be perfectly round. You can purchase a tool from a plumbing supply house that will help you straighten out a deformed nipple. If you have unions attached to the nipples, you can tap the union halves with a rubber mallet to get them to line up. If you are using couplers, you can't tap them, so instead insert a broom handle or other suitable tight-fitting implement into the nipple and carefully bend it into position. Be careful when bending the nipple that you don't deform it by making it out of round.

When plumbing flat plate collectors, it is important to have the supply piping attached to one of the ends of the bottom manifold and the hot return piping attached to the top manifold end on the opposite side of the collector. If you connect to two ports on one side, you will create uneven flow rate through the collector. If multiple collectors are used in an array, this same plumbing arrangement must be adhered to. You will want the array set up in a parallel design connecting the manifolds of each individual collector to make one large manifold. In most cases you will want to connect all

of the bottoms together and all of the tops together. This design will create an even flow rate throughout the array and will result in even heating. Even flow will also give the array maximum efficiency. It is also important to never exceed eight collectors in one array when using an internal manifold. This reduces damage that could be caused by expansion and contraction. If you have to, you can make an external manifold. However, this system will have slightly lower efficiency because of heat loss through the exterior pipes. If the collectors are in a saw tooth configuration, an external manifold is usually necessary. For residential applications, you would never want to plumb your array in series. You can end up with water that's too hot, as

FIGURE 8.12

Collector array in parallel with internal manifold

FIGURE 8.13

**Collector array in parallel with
external manifold**

well as decreased efficiency for the collectors on the end of the array. You would only want to plumb the array in series if you needed very hot water for commercial applications.

FIGURE 8.14

Collector array in series

If the ridge of the roof runs east and west, a large flat surface will be facing south, which allows for easy collector mounting. If the ridge runs north and south, a saw-tooth array can be installed (see Figure 8.15). A saw-tooth array uses more piping and is more difficult to install than an array that is parallel to the roof, but will work well if properly installed. For saw-tooth configurations, the distance between collectors is critical. If the collectors are placed too close together, they can shade each other. You need to make sure that the collector in the front will not shade the one behind during the winter solstice. Use the chart in the siting section to determine the angle of the sun during the winter solstice.

Some people build a catwalk several feet below the collector array to facilitate easy access to the array. This may be desirable in high snow areas, but is usually not necessary. Typically, snow slides off collectors quickly, especially for heating systems where the collectors are mounted at a steep angle to maximize the winter sun. After a wet sticky snowfall, the snow can freeze to the collectors, especially if the temperature drops significantly after the snowfall. A catwalk can make scraping the collectors a lot easier. This is not absolutely necessary because even ice will melt off the collectors when the sun comes out, but one can get a head start on heating if you manually

FIGURE 8.15
Saw-tooth array

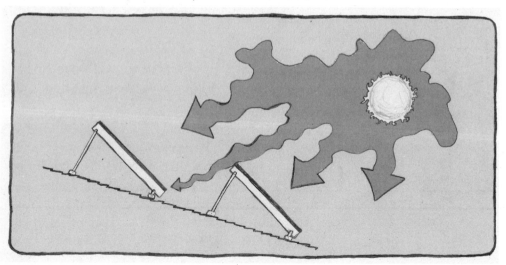

scrape off most of the snow early in the day after a snowfall. If a catwalk is installed, make sure it is located well below the bottom of the collectors to avoid snow buildup. Remember that snow will pile up at the base of the collectors as it slides off them.

The feed and return pipe runs to and from the collectors typically have to penetrate the roof and extend into the attic. Once in the attic, the pipes can be run to a convenient place where they can then extend down to the basement or mechanical room where the storage tank is to be located. The

best way to penetrate the roof is to use a roof boot. These items are commonly used to make a waterproof roof penetration for sewer vent pipes which are similarly sized to insulated solar piping. (An insulated three-quarter-inch or one-inch copper pipe will be three to four inches in diameter when insulated.) The roof boot is usually made of an aluminum plate with a neoprene collar mounted in the center of the plate. The insulated pipe fits tightly through the neoprene collar and requires no additional sealing. Some roof boots are made entirely of plastic. You will need one roof boot for each pipe. Typical roof boots are self-adjusting for any standard roof pitch. They are available at most lumber yards. Be sure to install them properly by fitting the top of the roof boot below the upper shingles and over the shingles below it. Use liberal amounts of roofing cement when installing roof boots.

Ground Rack Mounting

When there is not an appropriate roof to mount the collectors on, a ground rack mounting system can be used. Solar thermal collectors, whether flat-plate, ICS, evacuated tube, or concentrating, all have ridged frames. These frames can be bolted to a framework, result-

FIGURE 8.16
Roof boot

ing in a safe, durable, and long-lasting installation. The framework can be constructed out of wood or metal. The frame needs to be strong enough to withstand winds of over 100 miles per hour and needs to be securely anchored to the ground.

It is extremely rare to use ground mounts for any type of solar water heating or space heating system except closed-loop antifreeze systems. But when using the closed-loop system, ground-mounted arrays are very popular and offer a major advantage over other systems in offering a greatly expanded number of collector mounting options. The location of the ground mount rack should be identified using the criterion specified in the previous section on siting the collectors. You want to locate the rack as close to the home as possible to minimize the length of the pipe runs. Note that the piping between the collector array and the home will most probably be buried, so note any obstructions that would interfere with trenching, like a driveway or buried power lines. Always know what may be buried in the yard before you dig.

If you live in an area where snow accumulates, the rack should be designed so the bottom of the collectors is raised

above the ground. The bottom of the collectors should be at least twice as high above the ground as the deepest snow you would experience during an average winter. This height is needed because snow will slide off the collectors and make a pile at the base of the

FIGURE 8.17
Ground mounted collectors on concrete pillars

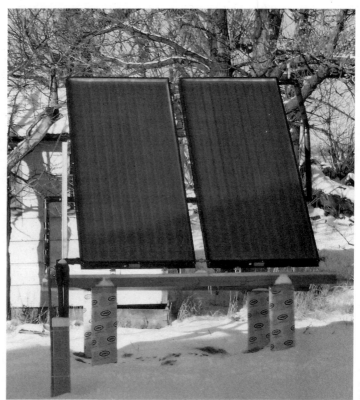

Bob Ramlow

array. You may still need to go out and compact the snow after a heavy snowfall, but this would be a rare occasion.

Ground mount collector racks can be built out of wood or metal. Because the collector array is subject to potentially high winds, the rack must be constructed to withstand the highest winds that normally could occur in your area. In areas subject to possible high winds, like coastal or mountain- ous areas, concrete footings should be provided to anchor the rack to the ground. Footings should extend below the frost-line in your locale. Concrete footings should be placed on undisturbed soil. The footing piers should have a framing member embedded in the cement or have an anchor bolt embedded at least four inches into the top of the pier. In areas where high winds are not experienced, wooden piers can be sunk into the ground below the average frost depth. Treated four-by-four wood works well.

FIGURE 8.18

Installing collectors on a ground mount

Midwest Renewable Energy Association

All collector manufacturers have collector mounting systems. These mounting systems include tilt mounting hardware or tilt racks. This hardware allows for easy collector mounting on any pitched roof or flat surface. The best way to ground mount solar collectors is to create two parallel rails and fasten the flat roof mounting system to those rails. The distance between the rails is determined by the specifications provided by the collector manufacturer. This would be the distance between the front feet and the rear legs. Parallel rows can be made of treated wood or metal rails.

The ground mount racks should be strongly built with adequate cross-bracing. Racks can be constructed of angle iron, galvanized pipe, or wood. Remember to isolate different types of metal with rubber, neoprene, or other non-conductive washers to minimize galvanic reactions between dissimilar metals. Aluminum angle iron is an excellent material to construct the racks from. If galvanized metal is used, be sure to properly paint all surfaces, especially the cut ends. Likewise, all wooden racks should be painted to extend their lifespan. Remember that these systems will last the life of the home they are servicing, so build accordingly.

I have had several customers incorporate a ground mount into a garden shed or storage shed. One customer used the area under the array to store firewood. Another used the building as a playhouse for their kids. In these cases, an actual roof was built on the rack and the collectors were mounted flush on the roof with standard collector mounting hardware. A roof was also built on the back side of the array to make an A-frame type structure. Add two end walls and presto, you have a shed for little more cost than that of the rack. Be creative!

AIR COLLECTOR MOUNTING

Because air-type solar heating systems are only intended to operate during the winter months when the sun is low in the sky, the collectors should always be mounted at least at a 60-degree angle or greater. Most air-type collectors are actually mounted at a 90-degree angle. Wall mounting has been most popular. The vertical mounting of the collectors takes advantage of snow to reflect additional solar energy into the collector for increased efficiency.

Mounting the Heat Exchanger

At any time during the installation process you can get the solar storage tank placed and start the plumbing of the heat exchange module. If you are using a pre-packaged heat exchange module, this part of your installation could be as easy as setting the module on your storage tank and tightening a few unions. If you are making your own heat exchange module with individual components, make sure your gauges and valves are visible and easily accessible.

Tube-in-shell heat exchangers can be mounted to the storage tank in several ways. The storage tank is typically an electric water heater. The elements are not used.

The first step is to set the storage tank in its appropriate location. If possible, it is best to elevate the tank by six inches. I suggest cutting three pieces of three- or four-inch PVC pipe six inches long. Set the storage tank on these new legs. If the storage tank has legs, place the new legs over the tank legs.

Using a new electric water heater (or one in very good condition) as a solar storage tank is a common practice. I use them most of the time because they are easily available, of high quality,

and are a good value. These tanks usually have two or three ports on the top — a hot port which is open to the top of the tank, a cold port which has a dip tube fitted to it, and a temperature and pressure relief valve port. If this valve port is not on the top, it will be on the side near the top. A drain valve with a boiler drain is at the bottom of the tank, and one or two electric elements are screwed into the side of the tank. All the ports have a three-quarter-inch national pipe thread (NPT) except the electric element ports which are one-inch machine thread.

There are two ways to retrofit an electric water heater into a solar storage tank. The first method uses the third top port and the drain port of the storage tank as connection ports. In this case the elements are left in the tank and are not used. The third port on the top of most electric water heaters has the temperature and pressure relief valve installed in it. You must remove this valve and install a brass nipple in this port. You can transition to copper pipe anytime. Install a tee on top of the brass nipple. Extend that pipe up above the tee four inches and reinstall the relief valve there. On the side branch of the nipple, run this pipe over to the side of the tank and down to the top port of the heat exchanger. Install an automatic air vent on this line. The lower water port of the heat exchanger can be plumbed to the tank drain port. Remove the drain from the tank and install a brass nipple. Install a tee on the nipple and install the drain on one leg of the tee. Run a pipe from the other leg of the tee to the bottom port of the heat exchanger.

If the tank has the valve port on the side of the tank, you can use method two when attaching the heat exchanger to the tank. This second method uses the one-inch machine-

FIGURE 8.20

Method two

FIGURE 8.19

Method one

FIGURE 8.22

Storage tank with vacuum breaker

threaded electric heating element ports. If you remove an element, you can see that the threads on the element are not tapered, and at the end of the threads there is a flat faceplate where an O-ring or gasket is fitted. When the element is screwed tightly into a water heater, the faceplate comes into contact with the water heater flange and the O-ring seals the fitting to the tank.

FIGURE 8.21

Double-pumped configuration

To make this port useable with tapered pipe thread, a modified 1" x ¾" brass bushing is used. I have a machine shop modify a standard 1" x ¾" brass bushing by turning the end of the nut flange flat on the inside where the threads taper into the nut. This creates a flat surface where an O-ring can be fitted, just like the element is set up for. With this modified bushing screwed into the tank, you have a new port at a perfect location to attach the top pipe from a thermosiphon or pumped heat exchanger. I use the drain port to attach the bottom of the heat exchanger to the tank. Remove the original tank drain and install a ¾" x 3" brass nipple in the port. I then install a ¾" x ¾" x ¾" brass tee on the nipple. On one branch of the tee I install a ¾" brass boiler drain. The other port goes to the heat exchanger.

If you are placing a second pump on the system, you will need to install that prior to mounting the heat exchanger. Figure 8.21 provides the location of the secondary pump. Ensure that this is a small pump to maintain stratification of the tank.

If you are using a plastic-cased fiberglass tank, you will need to install a vacuum breaker on the system, eliminating one of the ports on the top of the tank. You will need to install the cold inlet on a brass tee near the bottom of the heat exchanger. Figure 8.22 shows the new location of the cold inlet.

Plumbing the System

Many of the problems encountered during the installation of a solar water heater occur because of poor work-manship while plumbing. The most common of these is due to neglect while soldering the copper tubing. A soldered copper joint can last over 100 years if done correctly. If this is your first time soldering copper I would sug-gest that you practice a couple of times before working on your system. I should also stress that the instructions on sol-dering copper tubing must be adhered to for the joint to seal and last.

When retrofitting a home with solar, often the most difficult part is trying to find a way to get the pipes from the collector to the solar storage tank without having to tear out some sheetrock. There are a number of options for you to consider before that becomes necessary. I always start by looking for some sort of chase or recess in the wall that leads directly from the utility room to the attic. Often a home will have an enclosed chimney with a little extra space or chase that

has been built around the main sewer vent. This is always the best place to start, but I would warn you that unless you can see all the way through, there might be obstructions. Sometimes it may appear that you can make it from top to bottom, but when you try to shove some copper pipe through, you realize that the space isn't big enough or that the floorboards butt tight against the chimney. If there isn't a chase, the next best thing to look for is closets that are stacked upon one another. Most people don't mind a couple of pipes that are hidden behind their clothes. If there is only one story from the attic to the utility room, this is usually your best bet and will some-times work on a two-story home if you get lucky. If you are completely adverse to the idea of opening your walls, the last option is to make an exterior pipe run. You should only consider this if you are willing to insu-late your pipes well, especially if you live in a cold climate. Exterior pipes can often be hidden behind a gutter drain in order to retain the aesthetics of the home. I have also seen installa-tions where the pipes were enclosed in a box made of the same materials as the siding. This is probably the best way to make an exterior pipe run

FIGURE 8.23
Flux

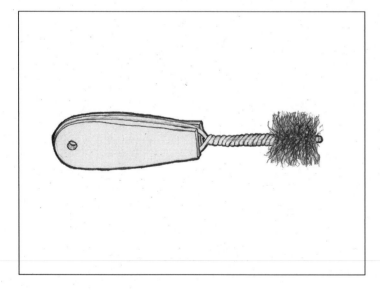

FIGURE 8.24

Wire brush

because you can add some extra insulation inside the enclosure.

I have already recommended that you make a simple line drawing of the pipe runs. This will give you an idea of the amount of copper tubing and the number and type of fittings you will need for the job. If you are going to be making a long pipe run, you will have to be able to slide the pipe in from either end and will need the space to do so.

Installing and insulating the pipes is the most time consuming part of most installations. The solar loop piping and pipe insulation costs can add up fast as well. The best location for the pipe runs is a trade-off of pipe run length and ease of installation.

Once you have determined the location of the pipe runs, you need to gather all the necessary materials:

1. Torch. Torches come in a variety of models, but most of the newer ones have a built-in igniter. With this type you only need to turn on the gas and click the trigger. For the older models you will have to ignite the flame manually.

2. Gas. The most common type of gas used is propane. Propane works fine for soldering, but you can use acetylene or Mapp gas if

you already have them available. They tend to burn hotter than propane and will cut down the amount of time you spend heating the copper. However, they increase the risk of overheating. The safest bet is to use propane.

3. Solder. Ensure that you are using a lead-free solder. Even though potable water may not be traveling through the solar loop of the system, it is never a good idea to bring lead into your home. I would recommend using the 95/5 lead-free solder that comes on a roll. It is composed of 95 percent tin and 5 percent antimony. You can also use the solder that is 100 percent tin, but the 95/5 is slightly stronger. Either type is suitable. I also wouldn't use a paste solder or a flux and solder mixed together, as your joints will not seal as well.

4. Flux.

5. Wire brush.

6. Grit cloth. Don't use emery cloth. Many types of emery cloth contain oil that may degrade the joint. If you plan to use emery cloth, ensure that it is oil-free.

7. Wet rag.

FIGURE 8.25
Pipe cutter

8. Gloves.

9. Measuring tape.

10. Pencil or marker.

11. Insulated heat mat or small sheet of metal.

12. Copper pipe and fittings.

13. Pipe hangers and fasteners.

14. Pipe cutter or hacksaw.

Once you have everything you need together, you can begin to assemble the pipe runs. I would suggest starting from one end, either top or bottom, and working the other way. If there are

FIGURE 8.26
Cutting the tubing

FIGURE 8.27

Separating the tubing

two of you working together, make sure you will be meeting in the same place. Since you already have the pipe runs planned out, you know where the fittings will need to go. Follow the next steps to ensure well-sealed joints.

STEP ONE: MEASURING AND CUTTING THE COPPER TUBING

From your starting point, measure the distance to the next fitting. You can either hold up the copper pipe and mark the distance or you can gauge it with a tape measure. Most likely, you will use a combination of both. With either method, be sure to take into account the length inside the fitting.

You have a couple of options for cutting the tubing. I would recommend using a pipe cutter like the one seen in Figure 8.25. You will need to place the circular blade at the appropriate length and hand-tighten it to the tubing. Spin the pipe cutter around a couple of times until you start to feel it loosen. Retighten the blade and repeat until the tube is cut. I would recommend that when using a pipe cutter you lead with the wheels instead of the blade. You can also use a hacksaw or a reciprocating saw with a metal-cutting blade, but I would discourage these methods because they do not always make a square edge and often leave metal burrs and chips on the end of the tubing. If you cut this way, you will want to use a metal file to clean off all the burrs.

I have found that the quickest way to install the tubing is to cut and dry-fit all the pieces together first, and then go back and solder them all together. If you are working in a tight space like an attic, you may want to work through all the steps instead of dry-fitting all the pieces.

STEP TWO: CLEAN THE TUBING AND FITTINGS

Thoroughly scrub the outside of the pipe and the inside of the fitting with either a wire brush or grit cloth. I typically use the wire brush on the inside of the fitting and the grit cloth on the outside of the pipe. Either one will work fine, but using both makes it a little bit easier. Cleaning the copper removes the oxidation and makes for a pure surface for the solder to adhere to. If you don't properly clean the copper, you will undoubtedly have leaks. The most common reason for leaky joints is skimping on this step. If you are soldering a lot of joints, you can purchase a wire brush that will attach to an electric drill to save some time. You will also want to scrape away any

FIGURE 8.28

Removing the burrs from the inside of the pipe

FIGURE 8.29

Cleaning the fitting with a wire brush

FIGURE 8.30
Cleaning the tubing with grit cloth

FIGURE 8.31
Applying flux to the tubing

burrs that have attached to the inside of the pipe.

STEP THREE: APPLYING FLUX

You cannot solder without flux. The solder won't adhere and will simply slide off the pipe. Most flux comes with a small fine-bristled brush. This is the best way to apply the flux to the copper. You will want to apply flux to both the outside of the tubing and inside of the fitting. Brush flux on all of the copper that will be inside the joint. You only need to cover the copper completely. You don't want gobs of excess flux remaining inside the pipe. Flux is acidic and may degrade your solar fluid if too much is left in the tube. Once the flux has been applied, slide the tube into the fitting.

STEP FOUR: APPLY HEAT TO THE JOINT

Ignite your torch and use it to apply heat to the joint and tube. You will want to heat the entire fitting and part of the tubing, hitting the flux directly with the flame. Try not to heat the place where the fitting overlaps the tube. When the flux starts to sizzle and boil, you know that your joint is hot enough. This process usually takes about 15 to 30 seconds when using propane. Don't overheat the joint. If the joint gets too hot, it may burn the flux and the solder will not adhere, or the burnt flux will block the flow of the solder.

You need to use caution when using a propane torch in tight areas. Often pipe runs will go through attics and crawlspaces without much room to maneuver. If you have to apply heat to a joint that is right next to a stud or something else flammable, be sure to place an insulated heat pad or a sheet of metal behind the fitting. Some insulated heat pads will have eyelets so you can hang them on a nail to keep your hands free.

STEP FIVE: APPLYING THE SOLDER

Unravel a couple of inches of solder from the roll and have it ready before you start applying the heat to the joint. You will use about the same amount of solder as the size of your pipe. For instance, for a half-inch pipe, you will use half an inch of solder. Next, take your roll of solder and touch the end to where the tubing meets the fitting. Direct the solder to the opposite side of the pipe than you are applying the heat to. When you apply the solder to the side opposite the heat, you know that the near side

FIGURE 8.32
Applying flux to the fitting

FIGURE 8.33
Heating the fitting

FIGURE 8.34

Applying solder to the joint opposite the location of the torch

will be hot enough. If you start applying the solder to the side near the flame, you will not be sure that the opposite side has reached the appropriate temperature and solder may not fill that space. If your joint is hot enough, you will see the solder melt and be sucked into the joint. Even if the joint is placed vertically, you will see the heat's capillary action draw the solder upwards. Once the solder starts flowing into the joint, you should take the heat away from the joint so as to not overheat anything.

STEP SIX: DON'T TOUCH IT

If the joint is rattled too much while it is still hot, the solder may not cool evenly, which may lead to leaks in the pipe run. Some plumbers will use a wet rag to wipe the joint while it is still cooling, but this is not necessary. You can still wipe excess flux from the joint once it has cooled, which will prevent the greening of the pipe over time.

STEP SEVEN: SECURE THE TUBING

It is important to securely attach all your piping to a wall or ceiling. A popular method professional plumbers use is to anchor a channel to the wall and use proprietary clamps that fit into the channel and clamp to the pipe. These

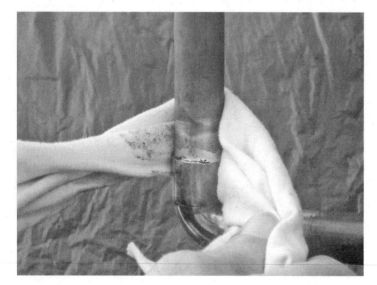

FIGURE 8.35

Wiping excess flux off the pipe

pipe-hanging systems make for a clean and solid installation. You can also use bell hangers or C-shaped clamps that nail or screw to the wall. It is important not to attach a circulating pump, especially a photovoltaic-powered pump, directly to an interior or a hollow wall because the vibrations generated by the pump will amplify in the wall and can become a nuisance. When stabilizing the piping near a pump, be sure to use rubber isolation clamps to minimize the noise.

When installing the pipe runs from the heat exchange area to the collectors, the piping needs to be supported every five feet of horizontal run. Many brands and types of pipe hangers will work just fine for this task. This is most important in drainback and draindown systems where any sags in the pipes could allow water to settle in the sags, freeze, and burst the pipe where the pipe runs through an uninsulated area. Another potential problem is that water collecting in sags can inhibit the quick drainage needed with these types of systems.

With ground-mounted arrays, the insulated piping is usually run underground from the array to the building. The trench should be at least 18 inches deep, and 36 inches deep under drive-ways in northern climates where the ground freezes solid during the winter. An important consideration when burying copper pipes is to compensate for the expansion and contraction of the piping. Copper piping will expand and contract considerably when it goes through its daily heating and cooling cycle, especially in cold climates where the temperature of the pipes can vary by over 200 degrees each day. When pipes are buried, tightly packed ground can prevent them from expanding when they warm. This can cause buckling of the

FIGURE 8.36

Attaching the pipe to the bell hanger

pipes and eventually cause leaks to form. The phenomenon is especially evident in long straight pipe runs.

There are a number of things you can do to minimize potential problems. The most important consideration is to use type K soft copper pipe for all buried piping. This type of copper pipe is relatively flexible and is somewhat self-compensating. Another advantage of soft copper pipe is that it comes in long rolls, usually 60- or 100-foot continuous lengths, which minimizes the number of joints that will be buried. It can also be bent, which also

FIGURE 8.37

Insulation mitered at the corner

minimizes joints by eliminating the need for elbow fittings. The best way to protect this pipe is to place it inside a PVC drainpipe. I like to use a four inch PVC pipe for each pipe. The four inch size is easy to slide over the pipe after it has been insulated and provides plenty of room for expansion and contraction. Some contractors prefer to place both insulated pipes inside a single larger PVC pipe. A six-inch inside diameter (ID) pipe is a tight fit. Be sure to glue all joints to keep them waterproof. If an elbow is needed in the underground pipe run you can make the job of installing the PVC elbow much easier by cutting the elbow in half lengthwise with a hacksaw, placing the halves together over the bent pipe, and then gluing the PVC elbow back together and to the straight lengths on either side. I also like to stuff the insulated copper pipe into the PVC pipe and glue it all together before dropping the whole thing into the trench.

STEP EIGHT: INSULATE YOUR PIPES

All pipes should be insulated. Exterior pipe runs should be especially well-insulated. Use a type of pipe insulation that will stay flexible after being exposed to the high temperatures experienced with solar water heaters. The best insulation I have found for this specific application is HT-Armaflex. This product only needs to be used on the hot or supply pipe, while regular Armaflex can be used on the cool or return pipe. This foam insulation will remain soft and cushion the piping when it expands and contracts. When installing pipe insulation, be sure to miter your corners to ensure insulation thickness over the joint. When insulating a tee in the pipe, cut a notch out of the long piece at the joint and cut the end that butts against the notch into a point to fill the gap.

Testing the System

Once the whole system is plumbed, it is time to check for leaks. There are two ways to check for leaks: one uses air pressure, and the other uses water pressure. Using air pressure has two advantages: no mess, and dry joints. To pressure test a system with air, you need to attach a fitting to the system to allow you to add air pressure to the piping loop. I use a bushing that has a Schrader valve installed on it. You take one of the boiler drains out of the solar loop and screw the bushing/valve into that fitting. Then take your

FIGURE 8.38
Insulation cut for a tee in the pipe

portable compressor and start filling the system with air. You can monitor your progress by watching the pressure gauge. Pressurize the system to at least 50 pounds per square inch. Once it's pressurized, watch the gauge to see if the pressure is dropping. If there is a big leak, the pressure will drop noticeably right away. Note that if you are checking during the daytime, the pressure will fluctuate because the sun on the collectors will heat the air in the system, causing the pressure to rise, and when the sun leaves the collector or goes behind a cloud the pressure will drop a little. If you have a leak and you have plenty of pressure in the system, the air escaping will create an audible noise, usually a whistling sound. Go over the whole system listening for a noise.

During the soldering process, you will have noticed that the solder filled joints in a predictable and consistent manner. If you have a fitting that did not seem to take the solder properly, it is a good idea to not insulate that joint till after you pressure test the system. If you have a ground mount that has buried piping, you should do the pressure test before you back-fill the trench.

You can pressure test the system with water too. The advantage of using water is that you should rinse the system anyway before you put the solar fluid in. By rinsing the pipes, you remove any leftover flux, which can degrade the solar fluid. The trouble with using water is that if you have a leak, the water may damage the building, especially if the installation is a retrofit and the building is finished. It is very important to have the collectors covered when using water to test the system, so please read the cautions outlined in the section on charging a system in the following chapter.

After the system is pressure tested, you can proceed to charge it with solar fluid as outlined in the section on charging a system. Make sure to remove the handles from the charging valves and hang them in a convenient but out-of-the-way location, and install caps on all drain valves.

Installing Solar Pool Heating Systems

Because solar pool heating systems are so common and mainstream, manufacturers offer very detailed installation instructions. Please read your instructions carefully before you begin. These instructions contain many tips that will not only save installation time but additional costs as well. Following these instructions will also help ensure proper performance. All the larger manufacturers also offer technical assistance for proper system design and component choices.

When installing these systems' collectors and piping, it is important to mount everything to facilitate proper drainage. The collectors must be angled so they drain, and all the piping should be mounted to drain. Always follow manufacturers' instructions when installing these systems.

When you order your collectors, you can always order the balance of the system's parts from the same vendor. They often come as a kit. Use black schedule-40 PVC pipe and black schedule-40 PVC fittings if possible for all your piping. Make sure you get solid PVC pipe, not foam core PVC. If you can't get black PVC, install white schedule-40 PVC pipe and let it age in the sun for a year or clean the pipe with PVC joint cleaner before attempting to paint it. Black piping can significantly improve the performance of your solar heating system. Anchor the PVC piping every five feet on horizontal runs and every ten feet on vertical runs. Make sure the clamps you use on the PVC pipe are oversized to allow the pipe to expand and contract.

Installing Radiant Floor Tubing

Radiant floor tubing can be installed in either a dump system or a system with water storage. Both types will follow similar rules for installation.

TUBING LAYOUT PATTERNS

The size of the tubing that will be used in a solar radiant floor heating system will determine the length of each tubing circuit. The smaller the tubing, the shorter the length of the

circuit. As heating fluid enters the tubing and flows through the circuit, the fluid cools down. Therefore, introducing the warmest heating fluid where the heat is most needed can maximize the heating system. This is typically along outside walls. Tubing spacing will also affect the performance of a radiant floor heating system. The closer the tube spacing, the more heat will be given off. Tubing is usually spaced closer together along outside walls, where more heat loss occurs, and farther apart in the center of a building.

FIGURE 8.39

Single wall serpentine

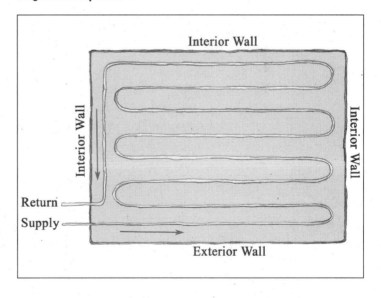

There are four basic patterns that can be used, depending on how many outside walls, if any, are in the area to be heated by a circuit. They all follow the same basic principle: send the warmest fluid to the locations with the most heat lost first. This will create even heating throughout the area.

SINGLE WALL SERPENTINE

When one outside wall represents the major heat loss of a circuit, the single outside wall serpentine design is recommended. As shown in the diagram, the hottest heating fluid is directed to tubing laid along this outside wall where the heat is needed most. The tubing then spirals to the area where heat is needed least. The tubing spacing should be closer where the heat loss is the greatest, and can be wider for the majority of the circuit.

DOUBLE WALL SERPENTINE

When two adjacent outside walls represent the major heat loss of a circuit, this pattern is recommended.

TRIPLE WALL SERPENTINE

When three adjacent outside walls represent the major heat loss of a circuit, this pattern is recommended.

COUNTERFLOW PATTERN

Use this pattern when the heat loss of the circuit is evenly distributed throughout the circuit.

PEX TUBING

The first modern radiant floor heating systems used copper or steel pipes, which were located beneath the floor, usually in concrete. Hot water was circulated through these pipes. These systems were very problematic because the pipes tended to break, causing leaks. The pipes broke because the metal expanded and contracted

FIGURE 8.40

Double wall serpentine

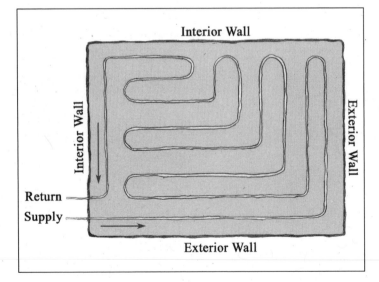

throughout the heating and cooling cycles. The pipes expanded and contracted more than did the surrounding concrete, resulting in great stress being placed on the piping. Eventually the pipes broke from the stress. Another problem arose when the concrete cracked. The metal pipes were not flexible, so they tended to crack along with the concrete. To make matters even worse, concrete corrodes copper.

Pex tubing is technically a cross-linked polyethylene. This product was developed in Northern Europe specifically to meet the special demands associated with radiant floor heating. Heating costs in those areas are double or triple what they are here in the USA, and radiant floor heating systems were known to use the least amount of fuel to heat a building. So a reliable tube was needed to do this job, and Pex tubing has stood the test of time. Pex tubing is flexible but very strong. It is nearly inert and does not corrode. It can also withstand moderately hot temperatures without deteriorating. It is a nearly perfect product for use in this type of solar heating system.

The advent of Pex tubing has allowed the advancement of high-mass solar heating systems.

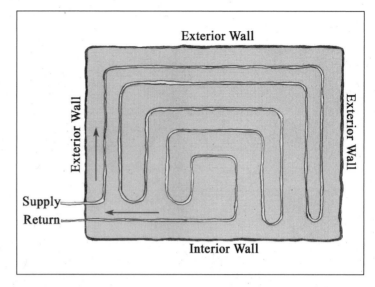

FIGURE 8.41

Triple wall serpentine

FIGURE 8.42

Counterflow spiral pattern

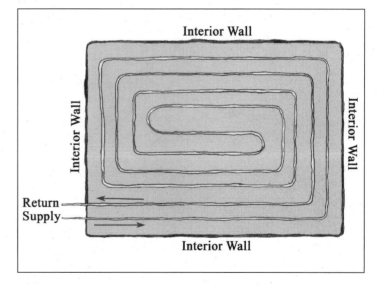

I suggest using a minimum size of half-inch ID tubing for the Pex tubing loops in the floor or sand bed. Some people use five-eighth-inch ID tubing. Both sizes work well. There will be slightly less resistance to flow in the larger tubing. You will need between 1 and 1.25 feet of tubing for every square foot of floor area.

TIPS FOR INSTALLING PEX TUBING

Pex tubing comes coiled from the supplier. While this tubing is flexible, it can be stiff when cold. Always keep the tubing in a warm place for at least six hours before uncoiling. When uncoiling it, let the tubing set in the sun for a while before placing it in its final location. This will soften up the tubing and make it easier to place.

When removing the tubing from the box, make sure you do not cut the tubing. The tubing cannot be used directly from the box. You must uncoil the whole roll before placing the tubing. If you try to just take the tubing off the coil without unwinding it, it will not lay flat and will create big spirals of tubing which will get all tangled up.

Put a piece of duct tape over the ends of the tubing to keep dirt from getting into the tube. The glue from any tape will deteriorate the tubing, so you will have to cut the ends that were taped. Remember to account for this when locating the ends of the tubing in preparation for attachment to the tubing manifolds. Never use tape to hold the tubing in place.

The simplest method of uncoiling is to have someone hold the end of the tubing, which is on the outside of the coil. Then roll the coil out on the ground as if it were a wheel. Most coils are at least 300 feet long, so you must have at least that much open space to work in. While rolling the coil, you must keep it from getting away from you. If it falls to the side or some parts of the coil begin to hang from the side, a tangle could happen.

Another easy method of uncoiling is to have one person take the end of the tubing that is on the outside of the coil. The second person holds the coil vertically like a tire on a car with his arms clasped in the middle. The person with the end then walks slowly away from the person with the coil. The coil person must help the coil unroll by flipping the coil to facilitate uncoiling. This takes a little practice, but can go quite fast.

You can also use an uncoiler. This is like a lazy Susan, a platform that turns horizontally. Place the coil on the lazy Susan and while one person takes the end of the tubing that is on the outside of the coil and begins to pull, the other person holds the coil on the uncoiler and helps the tubing come off without tangling.

The warmer it is outside, the easier it will be to install the tubing. The warmth makes the tubing more flexible. It is also helpful if it is sunny, for the same reason. I often leave the tubing uncoiled and in the sun for an hour to allow it to soften up. A few hours in the sun won't damage the tubing, but prolonged exposure to the sun is not recommended. The tubing should not be exposed to the sun for over 30 days. Consult the manufacturer of your tubing for their recommendations. It is usually not a problem for the bulk of the tubing because it is usually covered with dirt or concrete soon after layout. If, by chance, the tubing will be laid out but not covered for some time, cover it with tarps or anything that will keep the sun off. Take care to cover the tubing ends that protrude from the slab if they will be exposed to the sun for any length of time.

There is a limit to the length of tubing used in a circuit. If the tube length is too long, the hot fluid passing through the tube will cool off before it gets to the end of the circuit, and heating will not take place evenly. For half-inch and five-eighth-inch tubing, the maximum length for each circuit is 300 feet. Tubing circuits can be shorter than 300 feet. Many manufacturers package Pex tubing in 300-foot rolls. It is easiest if all circuits are relatively the same length, which makes balancing the system automatic. This is especially true in high-mass systems. The goal of balancing this part of the system is to have the temperature of the solar fluid where it exits the sand bed the same temperature in all the circuits. When the tubing circuits are all the same length, they will naturally all perform the same way. Shorter circuits will lose less heat because they are shorter, so the exiting solar fluid will be hotter. If the circuits are not balanced, uneven heating could occur. If all circuits cannot be the same length, ball valves can be used to adjust the flow of solar fluid through each circuit.

Your system designer should calculate the best distance between tubing for you. If your specifications call for 12 inch spacing, you will need 1 linear foot of tubing for each square foot of area. A 300-foot circuit will require 300 square feet of area. If the specifications call for 9-inch spacing, a 300-foot circuit will require 222 square feet of area. If the total area to be covered by the gridwork of tubing is 900 square feet, and the tube spacing is 1 foot, you need to divide the total area into three equal sections, one section for each circuit. A point to remember here is that you will have less than the full length of tubing to work with in most sections because some of the tubing length will be used up in going to and from the manifold location to the area to be covered with the gridwork. Be sure to space these tubing runs the same as all the rest of the tubing.

FIGURE 8.43

Radiant floor tubing attached to rebar in concrete slab

Once you get the total area divided into sections, it is time to start laying tubing. The surface where the tubing will lay should be prepared properly. The ground should be leveled at the proper level. Insulation is usually laid down first. A vapor barrier is also usually incorporated below the insulation. If the tubing is to be located in a slab, the total area on top of the insulation should be covered with reinforcing wire or rebar. Reinforcing wire comes in big rolls or 4' x 8' pieces. The wires usually form six-inch squares. This wire spacing can help with the tubing spacing. You can follow the wire to achieve a consistent tubing pattern. If using rebar, place the rebar to assist

FIGURE 8.44

Tube exiting the slab

Bob Ramlow

tubing placement. You can tie the tubing to the wire or rebar with nylon zip-ties or with short pieces of thin wire. Some manufacturers can provide you with either wires or zip-ties. Use a fastener every 12 to 18 inches, closer at bends. The tube needs to be firmly affixed to the wire or rebar to hold it in place when the concrete is poured. The tubing is full of air when the concrete is poured and it tends to float up into the slab. You want to minimize this and keep the tubing as close to the bottom of the slab as possible. Fasteners are cheap. Use lots.

Start where the tubing will exit the slab. Use a board with three-quarter-inch or larger holes drilled through it. The holes should be two inches apart and there should be two holes for each circuit drilled in a row. This board will help hold the tube ends in a neat row. Leave plenty of tubing protruding past the slab top to allow for connecting the tubing to the manifolds. Think about where the manifolds will be located and allow plenty of extra tube length. Remember that you will have to trim the ends of the tubing.

In order to protect the tubing where it exits the slab, a piece of electrical conduit should be slipped over the tubing and placed so the top of the conduit extends several inches above the top of the finished slab. Three-quarter-inch conduit works with half-inch tubing. If you are laying the tubing in a slab, use 90-degree elbows for this. The preformed elbow has a gentle sweep and holds the tubing in place as well as protecting it. Slide the elbow down the tube till one leg of the elbow is parallel with the ground and flat on the wire or rebar. The other leg of the elbow will be pointing straight up and should protrude through the top of the soon-to-be-poured slab. If you are placing the tube in a sand bed, use straight pieces of conduit about one foot long. The 90-degree bend the tube needs to make will be well below the slab, so you don't need the preformed elbow.

Start with both ends of one circuit fixed through the board through adjacent holes. I mark each tube end with an arrow indicating the direction of fluid flow, the hot arrow pointing down and the return arrow pointing up. Do not mark on the tube with a marker. Put the arrows on the tape and on the conduit where it will be above the slab top. This will be very important when it comes time to attach the tubing to the manifolds.

Now start to place the tubing for the first circuit. This circuit should be the farthest from the manifold. The feed end of the tube (from the hot manifold) should be placed nearest the exterior of the building and the return end of the tube toward the inside of the building right next to the feed tube. These tubes should run parallel to each other and be spaced apart at the designated distance. When you reach the area to be covered with this circuit, then lay out the tubing in the predetermined pattern. Remember to keep the hot tube closest to the outside of the building. Remember that because this circuit is farthest from the manifold location, the tubes have had to run through the area designated for other circuits. Because of this, the tubing will not quite fill the area that was designated for it. That is OK because when you lay out the rest of the circuits some of their area will be covered by the far away circuits as they make their way to and from the manifolds.

When you are done you will have a pair of tube ends for each circuit held neatly by the board with holes in it. You can adjust the placement of the board so it comes up near the header location. After the slab is poured and the building is closed in, you will

attach the manifolds to tubing and run the required piping.

CONSTRUCTING THE SAND BED FOR HIGH MASS SYSTEMS

What I am calling the sand bed is an insulated box filled with sand, which is buried beneath the floor. Exactly how this box is built will depend on how the building it serves will be constructed. Here are some examples. Please note that it will be up to the contractor to decide what will be the best method depending on site and climate. His experience in building in

FIGURE 8.45

Insulated box for high mass sand bed

your area should give you the most stable building. The important factor here is that the exact design of the sand box is flexible and should be fit into the design of the footings and slab of the building.

Consult the illustration of the high mass system detail in the description of that system to get a good impression of the order of the layers and their layout. (See also figure 6.4). If a radon mitigation system will be installed in the building, it should be installed beneath the sand box. A vapor barrier can be installed either above or below the insulation. Most contractors prefer to have the vapor barrier on the outside of the insulation.

INSTALLING PEX TUBING IN THE SAND BED

If you are placing the tubing in a sand bed you will not have any wire or rebar to attach the tubing to. Place the vapor barrier and insulation as described above. Next fill the sand bed to the depth of the tubing gridwork and level. In a sand bed, it is not as critical

FIGURE 8.46

Footing detail

to achieve a perfect grid of tubing (the sand is somewhat forgiving compared to a slab), but you still want to have a fairly even distribution of tubing. As you start to lay the tubing you can hold it in place with sand. It is helpful to have one person placing the tubing and another with a shovel plopping piles of sand on the tubing once it is in place. If the sand is damp, you will only need a shovelful every several feet. With dry sand, you may need a continuous covering. Wherever the tubing makes a bend, plenty of sand is usually needed to hold it in place. It takes some practice not to pull the tubing once it is laid. Even in warm conditions, the tubing seems to have a mind of its own and will want to curl up on you. Once you develop a successful method, it goes fast. Compact each layer of sand as it is added. If you have trouble keeping the bends in place, some contractors lay a piece of rebar down and attach the tubing to the rebar.

FIGURE 8.47

Installing tubing in a high mass sand bed

Marguerite Ramlow

Chapter 9

SYSTEM OPERATION
AND MAINTENANCE

~~~~~~→

Charging a
Closed-Loop System

CHARGING A SYSTEM is the process of adding the solar fluid to the closed loop and getting all the air out of that loop.

CAUTION!

All liquid-type solar collectors can produce steam! Extreme caution is advised whenever filling or draining closed-loop solar energy systems. Always cover solar panels when filling or draining a solar energy system during the day. Never open a closed-loop system when the panels are exposed to the sun. Never charge a system when the panels are exposed to the sun. This is no joke! Be careful. Failure to follow the above precautions could result in **severe burns** in addition to coating the inside of your collectors with gook (which is bad).

Please read through this whole procedure before attempting to charge your system.

List of materials needed to complete the task:

1. Charging pump

2. Three four-foot hoses with female hose fittings on each end of one and one end of the other two

3. Non-toxic glycol to fill the system

4. Distilled water to dilute glycol

5. Channel-lock pliers

6. Rags

7. Empty five-gallon pail

8. Pair of gloves

9. Trouble light

The charging pump should be a liquid transfer type, large enough to pump the solar fluid up to the top of the collector array. The greater the vertical distance between the charging ports and the top of the collector array, the larger the size of charging pump that is needed. I use a three-quarter horsepower jet pump and I can pump 40 feet vertical with no problem. A one-half horsepower pump will work up to 28 feet vertical. Many plumbers use this type of pump regularly. They are also relatively inexpensive to purchase. The pump needs to be outfitted with male hose fittings on both its inlet and outlet. Male hose fittings can be purchased at any good hardware store. You will need a male-hose-to-male-NPT thread adapter and perhaps a hex bushing to attach to the pump.

This pump will first have to fill the system with solar fluid by sucking pre-mixed solar fluid from a five-gallon pail. Once the system is full, the pump will then vigorously circulate solar fluid throughout the system, thereby flushing out all the microscopic air bubbles that can be difficult to purge any other way. Finally, the pump will be used to pressurize the system to 32 pounds per square inch at 60°F.

FILLING THE SYSTEM

Locate the charging ports in the plumbing header. The charging ports are the two boiler drains with the check-valve between them. Set up the charging pump near enough to the charging ports so that each of your lengths of hose can easily reach between. Set the empty five-gallon pail near the pump and charging ports. Attach one end of the hose with two male ends to the outlet of the charging pump and attach the other end to the charging port that the arrow on the check-valve is pointing to. This is the input valve. Fully open the input valve. Attach the second hose to the other charging valve (this is the drain valve) and put the other end of that hose into the pail. The end of this hose must reach the bottom of the pail. Fully open the charging port drain valve. Attach the last hose to the inlet of the charging pump and put the other end into the pail. This hose also must reach the bottom of the pail.

If there are any ball valves in your system, make sure they are all fully open.

Fill the pail with pre-mixed glycol and distilled water. (See the section on solar fluids in Chapter 4 for the proper dilutions.) You are now ready to begin charging the system with fluid. Make sure that your collectors are completely covered and out of the direct sun. Charging the system at night does not require panel covering, but panels must be covered under all other circumstances, even if it is a cloudy day. You also must be able to see the pressure gauge at all times while charging the system. If you cannot see the pressure gauge, have a helper stationed at the pressure gauge to give you readings at all times during the charging process.

If you have a self-priming pump, you are ready to start. If you do not have a self-priming pump, please prime it now by removing the hose from the outlet of the pump and filling the pump with some of the pre-mixed glycol and then replacing the hose. Firmly tighten all hoses with your channel-lock pliers. With one hand, grab the two hoses that are going into the pail. You

want to make sure that the ends of both hoses remain near the bottom of the pail. Never let go of these hoses while the pump is running and never leave the charging pump running unattended.

Turn on the charging pump. If the pump was properly primed, it should start pumping fluid from the pail into the closed-loop of your system. You can tell that things are going properly if the fluid level in the pail begins to go down. Make sure to keep hold of the hoses in the pail so they don't start flopping around. Carefully watch the pressure gauge. You want the pressure to stay between 10 and 25 pounds per square inch. You can control the pressure by adjusting the inlet filler valve. Close it down till you maintain the correct pressure.

Right after you turn on the pump, air will start to come out of the system through the drain hose that is in the pail. Because the end of this hose is at the bottom of the pail (below the fluid level), bubbles will begin to rise in the pail. As the level of fluid in the pail begins to drop, add more, trying to keep the pail at least half full at all times. Keep the empty pails you accumulate nearby in case you need to drain fluid from the system.

While the system is filling, you may hear various gurgling and bubbling sounds. This is normal. At a certain point, your system will be nearly full of fluid and devoid of air. You can tell because the bubbling in the pail subsides and the fluid level in the pail stops going down. Make sure that you are monitoring the pressure constantly. When the system becomes full, the pressure will want to go up. Adjust the inlet valve to maintain proper pressure. Continue to circulate fluid throughout the system with the charging pump.

If you find that as you begin to pump fluid into the system, the pressure rises quickly and keeps going up, while at the same time no air comes out of the drain port and hose, you should turn off the pump before the pressure gets above 50 pounds per square inch. Be careful at this point because pressurized fluid may flow back vigorously into the pail through the pump. If fluid will not circulate into the solar loop but the pressure builds quickly, your system may be plugged or it may have an air lock. First, check that all ball valves within the system are open. Next double-check that the plumbing runs have been hooked up properly. You should

be able to follow the pipe runs from the inlet port up and through the collectors and then back down through the heat exchanger and finally to the outlet port. If you find that all valves are open and the pipe runs are plumbed properly, then you have an air lock in the system. A good way to overcome an air lock is to first close the inlet port valve while leaving the drain valve fully open. Turn on the pump and slowly crack open the inlet valve. The pressure in the system will slowly rise. Allow the pressure to go to 60 pounds per square inch and then close the filler valve and shut off the pump. Make sure the drain valve remains fully open and always hold the drain hose firmly. Usually this pressure is enough to push any air bubbles through the system after a few moments and to allow the solar fluid to continue on through the system and back to the drain port. If there is still no circulation, try closing and opening all ball valves. Sometimes air bubbles can hang up in these valves. As a last resort, bring the pressure up to 75 pounds per square inch. The pressure relief valve will begin to open at this point. **Never bring the pressure above 75 pounds per square inch**, and always hang on to the drain hose firmly.

If you have an assistant, have him or her look over all the piping to check for leaks. Make sure to check the fittings between each collector (if you have more than one) and wherever there is a threaded fitting in the system. If you did everything right, there will be no leaks. If you do not have an assistant, you need to check it out yourself. To do it yourself, you must put the system under some pressure and you must turn off the pump. To do this, simultaneously close both filler and drain valves on the system. **Watch your pressure!** When both valves are closed, check the pressure. If it is above 23 pounds per square inch, turn off the pump and go check the system for leaks. If it is under 20 pounds per square inch, then slightly open the filler valve only. *Watch your pressure!* When the pressure reaches 25 pounds per square inch, close the valve and turn off the pump. Go check your system.

If you discover a leak, you will have to drain the fluid from the system and go back and solder the leaky joint or retighten the fitting. If you found no leaks, please continue with the charging process.

The next step is to continue to flush all air from the closed loop and, lastly, to pressurize the system. If you had to shut off the pump, turn it back on, and then slowly open both valves, opening the drain valve first. As always, watch your pressure to maintain 10 to 25 pounds per square inch pressure in the system. The pail must be over half full at this point and you will need a good light to observe the fluid in the pail. You should see the pressure in the system remaining constant, and a swirling action within the pail created from a fast stream of solar fluid emitting from the drain hose. The end of the drain hose has always remained near the bottom of the pail. You will notice that the color of the solar fluid is milky white, and that surges of air will often bubble out. You must observe carefully what is going on in the pail. Closely observe color changes and bubble size. You have to look carefully during the final stages of purging because near the end of the procedure the bubbles are very small and hard to see. Nonetheless, you need to get as many of those little bubbles out of the closed loop as possible.

After several minutes, the amount of bubbling will begin to subside and the milky color will fade. The milky color is caused by millions of microscopic bubbles. If you are installing a

Radiant Floor Systems

If you have a bypass pipe around the hot water heat exchanger, it should have a ball valve installed. Close that valve halfway. You should have a main heating circuit and a shunt circuit. Completely close the shunt loop valve, diverting all the flow through the heating system. If you have valves on each individual tubing circuit (recommended), leave two circuits open and close the rest. If your tubing is translucent, you can observe bubbles being flushed from the system. After those loops are clear of bubbles, open the next two circuits and then close the ones that were flushed. Work through the system till all are flushed. After the floor circuits are flushed, open the shunt loop valve and then close the floor valve. Pump till no more bubbles are observed. Finally, open all circuits again.

solar water heater without additional heating circuits, then proceed to the next paragraph. If you have a combination system with multiple load circuits, you now need to flush each individual circuit. During this procedure, always keep at least one circuit open at all times.

PRESSURIZING THE SYSTEM

I recommend that you continue to circulate fluid for at least five minutes from when the last bubbles were observed in the pail. Now slowly begin to close both filler and drain valves, maintaining at least ten pounds per square inch pressure. When both valves are closed, slowly crack open the filler valve to allow the pressure to rise to 32 pounds per square inch. This is most easily accomplished by allowing the pressure to rise to 35 pounds per square inch before shutting off the filler valve. The pressure gauge may bounce around a little while you are opening and closing valves. Turn off the pump. Let the pressure settle down. Then slowly crack the drain valve to release pressure down to 32 pounds per square inch at 60°F solar fluid temperature. If the solar fluid is cold, lower charging pressure to 30 pounds per square inch.

You are almost done. Have your rags handy and grab the channel-lock pliers. Tighten both charging valves very tightly. Remove the hoses from the charging valves. Install caps on the charging valves and tighten them securely. As you recall from previous chapters, liquids expand and contract with temperature changes. This will be reflected in pressure changes on your system gauge. At noon on a clear day in the middle of summer, you may observe pressure over 40 pounds per square inch but at night when it is 20°F below zero you may observe 15 pounds per square inch. But most of the time, you should observe pressure readings around 30 pounds per square inch. If you notice a consistent drop in pressure, you probably have a leak. Thoroughly check the system, paying special attention to non-soldered fittings. If you are having trouble locating a leak, try isolating individual circuits and observing pressure changes. After you've determined that the system is operating properly and with no leaks, remove the handles from the two charging ports as a safeguard against accidental opening. Hang the handles from a zip-tie on the pipe nearby. Tape the following information to the expansion tank: date of installation, servicing dealer or plumber, type of solar fluid, solar fluid concentration, and initial pressure. Clean up and you're done.

FINISHING UP

After you are done charging the system, you should clearly label all the main lines with identification tags. The tags should identify what the lines are and the direction of flow. You should also assemble a booklet about the system that contains the following information:

➤ Collector literature and warranty

➤ Tank information and warranty

➤ Pump information and warranty

➤ Description of system operation

➤ Controller information and warranty (if applicable)

➤ Solar fluid information

➤ Maintenance information

Controls and Power Sources

All solar water and space heating systems require a pump to circulate solar fluid through the solar loop, except for thermosiphon and ICS systems. These circulation pumps have to operate whenever there is solar energy to harvest, but cannot run during non-collection hours. If the pumps ran all the time, the solar collection system would become a heat dissipation system, especially at night when it is cold. Therefore, if the pump can't run all the time, there has to be some kind of brain to turn it on at the appropriate times, as well as turn it off when there is no longer solar energy to collect or when the storage is up to its maximum temperature.

There are basically two ways that these circulating pumps are controlled. The traditional method is to use an AC-powered pump that is controlled by a differential temperature controller. The other method is to use a DC-powered pump that runs directly off a photovoltaic module.

AC-POWERED CONTROLS

In AC-powered systems, the power to run the pump comes from our regular 120 volt AC home electrical system using utility-generated electricity. Of course, that electricity is "on" 24 hours a day, so if the pump is plugged into an outlet, it would run all the time, which you don't want happening. A differential temperature controller uses sensors that electronically measure temperatures at remote locations. In a solar water heating system, one sensor is placed to measure the average temperature of the storage, while another sensor is placed to measure the temperature of the collectors. The differential temperature controller compares the temperature of the storage and the temperature of the collectors. Whenever the collectors are hotter than the storage, the controller switches the pump on. The pump will continue to run till the temperature in the collectors nears the temperature in the storage.

A modern differential temperature controller uses a small microprocessor to analyze the information it gets from the sensors and a logic chip is used to decide when to do the switching. The controller is powered by 120 volt AC and is either hard-wired to a fused circuit in the building's AC load center, or is plugged into an outlet like any other household appliance. The microprocessor operates on low voltage, so there is a transformer in most controllers. Note that the sensors also operate on this low voltage. In these systems, the pump is 120 volt AC powered. In order to turn the pump on and off, a relay switch is also included in the controller.

There are numerous brands of differential temperature controllers on the market today. Most of them are designed to work for a number of different types of solar water heating systems. There are usually a number of settings that must be made to the controller when it is installed so it will function properly for your particular system. Many also include a high-limit function that turns off the pump when the storage reaches a pre-set temperature. Many controllers also have a recirculation mode, which is used for flooded systems. When the recirculation mode is enabled, the controller will turn the pump on when the collector is approaching freezing conditions. Warm water is then drawn out of the storage tank and circulated throughout the system, offering protection during occasional freezing episodes. When shopping for a differential controller, make sure it has the features you will need for the type of system you are installing. Also make sure to read the instructions that come with the controller, so you get the settings correct.

In the 1980s, many companies made their own controllers, which were often very simple devices. Today, most AC-powered systems use mass-produced controllers. Although today's controllers are very reliable and affordable, I must add that replacing controllers has been one of my most common service calls over the years. The problem is that this type of electronic device is very vulnerable to electrical discharges, caused primarily by lightning, but also by spikes from the electric utility. Most of the failures I've encountered were due to improper grounding of the solar energy system. If you suspect lightning damage, open the cover and look at the circuit board for burn marks or black blotches. Sometimes the mark is hard to find, but often it is big and obvious. Relays are sometimes burnt as well. The relay is usually enclosed in a plastic cover, which is often clear.

Sensors are typically quarter-inch-diameter by one-inch copper cylinders that have two wires coming out of one end. Sometimes one end is flattened. Two wires connect the sensor to the controller. Every controller requires a specific type of sensor to operate properly. Most controllers use sensors that are identified as 10K sensors. These can be identified by their black wires. Some older controllers use 5K sensors, which are identified by their white wires. I recommend that you use # 16

FIGURE 9.1

Flat bolt-on type sensor

FIGURE 9.2

Immersion type sensor

FIGURE 9.3
Screw-in type sensor

FIGURE 9.4

Well type sensor

stranded sensor wire that is double-insulated. When wiring the sensors, I also recommend soldering all electrical connections. The connections then should be sealed with heat-shrinkable tubing. Using wire nuts allows corrosion to form at the connection, causing inaccurate sensor readings. This is especially important for the collector sensor, which is exposed to a harsher environment than the tank sensor.

The sensor that is located at the collector is usually mounted on the pipe that exits the collector and delivers the hot solar fluid to the heat exchanger. The sensor should be located as close to the collector as possible. You may be able to actually slip a part of the sensor under the rubber gasket where the pipe exits the collector frame. The sensor is attached to this pipe with a stainless steel hose clamp. Some installers locate the sensor on the top pipe between the collectors (if multiple collectors are used) also using a stainless steel hose clamp. Either location is acceptable.

The tank sensor is typically mounted about one-quarter to one-third of the way up from the bottom of the storage tank. The sensor must make good contact with the side of the tank. Most solar storage tanks are insulated with foam, so the sensor must be mounted between the foam and the tank. Most tanks have an access door near the bottom of the tank. Some tanks actually have a stud welded to the side of the tank, allowing easy sensor mounting. If no stud is available, the sensor needs to be firmly pressed to the side of the tank. Often, a piece of foam insulation can be used to wedge between the existing insulation and the side of the tank jacket to hold the sensor firmly against the tank. This is a common practice when using an electric water heater as the storage tank.

DC-POWERED SYSTEMS

A very popular system design today utilizes a 12 volt DC circulating pump and a dedicated 12 volt DC photovoltaic module to power the pump. This type of system does not require a controller. The photovoltaic module is mounted near the thermal collectors, so the only time it produces electrical power is when the sun shines on it, which is also when solar thermal energy is being collected. At night, or during very cloudy conditions when the system should not be operating, the pump will not run because there is

no solar energy available for the photovoltaic module to make electricity. During low light conditions, the solar thermal collector's output is reduced and so is the photovoltaic module's. If there is enough solar energy available to create heat in the thermal collector, there is enough to make electricity in the photovoltaic module as well, but at a reduced rate. The pump will vary its speed in direct relation to the amount of solar energy available. This is a perfect match, as the pump speed exactly matches the circulation needs of the solar thermal collector.

FIGURE 9.5

Pool type sensor

Wire Nuts

Wire nuts are used to fasten two wires together and are commonly used in 120 volt AC electrical wiring. Wire nuts should not be used in low voltage wiring, especially on sensor wiring circuits. A sensor is a resistor. It has different resistance for each temperature. Corrosion on a wire is also resistance which may give false readings to the differential temperature controller. For these low voltage applications always solder the connections and use heat shrink tape to seal the connection. This is especially important when the sensor wire connection is exposed to the elements.

Grounding Collector Arrays

It is important to properly ground all solar thermal systems, but it is particularly important to ground AC-powered systems to protect the controller. Solar collectors can act like big static electricity absorbers, so when lightning strikes nearby, some of that electrical energy is caught by the collectors. The whole system is tied together by copper pipe, so that static can be routed through the system to the controller.

When choosing a 12 volt DC circulating pump, it is important to pick one that is designed to operate directly off a photovoltaic module. Brushless pumps are best. Most brush-type pumps will also work well, but the addition of a linear current booster in the circuit between the PV panel and the pump often increases the reliability of the circulating pump, as the booster will help the pump start properly.

With any electrical circuit, wire sizing is important. With low voltage circuits, it is very important to use the proper wire size for proper system

FIGURE 9.6

Exterior pump enclosed in weatherproof box

operation. There is a direct relation between system voltage, amp draw of the circuit, and length of the wire. In general, the lower the voltage of a particular circuit, the larger the wire size needed. If the wire is too small, not enough current will flow through, and in this case the pump will not operate properly. Please refer to the wire-sizing chart in Appendix A when sizing your wire from the photovoltaic module to the pump. The figures you will need when referring to the chart are the amp draw of your pump and the length of wire needed to connect the photovoltaic module to the pump.

The traditional location for the main circulating pump is near the heat exchanger and storage tank. This is still the case with AC-powered systems. With low voltage systems like photovoltaic-powered systems, it is common to locate the main circulating pump near the photovoltaic module and the thermal collectors. The main reason for this is to save money and resources by eliminating a long wire run. This wire run can get expensive because of the 12 volt DC circuit requiring large wire sizes. If the collectors are roof-mounted the pump can be located either in an accessible attic or near or behind the collectors in a ventilated enclosure to protect the pump from the elements. If the collectors are ground-mounted, the pump is usually mounted behind the collectors in a weatherproof and ventilated box. It is important that this box is ventilated in order to prevent overheating.

AC AND DC SYSTEMS COMPARED

Some types of solar water heating systems only work with AC-powered pumps. Draindown and drainback systems require high-head pumps which are not available in 12 volt DC at this time, so they are always AC-powered and use a controller. Any system that requires any type of motorized valve will also have to be AC-powered, but these are rare.

With AC-powered systems, the circulating pump runs at a constant speed because the power source remains constant. With photovoltaic-powered systems, the pump speed varies depending on the intensity of the sun. This difference in pump operating speed affects the overall performance of the system in a marginal way.

With AC-powered systems, when the pump starts in the morning it is running at full speed while the amount of available solar heat is minimal. If there is still some heat in the storage tank or if it is very cold outside, the tank can actually be cooled slightly till the sun's intensity becomes greater. This same situation can also happen during partly cloudy weather as clouds pass over the collectors. This could result in some inefficiency in overall performance. On the other hand, in AC-powered systems the system will turn off when the storage tank either reaches its high limit or when the tank is warmer than the collectors. This is a good thing in most cases. The only time this is not good is with closed-loop antifreeze systems. If the system isn't circulating under high heat situations, the solar fluid could get extremely hot, which could lead to its deteriorating. This would happen under low demand situations, like when the household is on vacation during the summer.

With DC-powered systems, when the pump starts in the morning it is running slow because the solar resource is low at that time. The result is that the pump speed exactly matches the output of the solar thermal collectors, resulting in optimum performance. This same situation can be observed under partly cloudy con-

When installing collectors, it is important to electrically bond all the collectors together, and to the copper piping, and then to a dedicated ground rod. A typical grounding system uses # 6 bare and stranded copper wire, which is screwed to each collector and then bonded to the system piping; it is then connected to a dedicated ground rod. When attaching the stranded ground wire to the collectors, you can just spread out the strands, place a stainless screw through the wire, and screw the wire tight against the collector. If the collector is painted, it is best to scrape the paint off where the ground wire is attached to the collector. The best grounding systems always use a continuous grounding wire, so it is best not to splice the wire, but use one long strand. Start at the collector farthest away from the ground rod, attach the wire to each panel as the wire travels to the other end of the array, attach the wire to the piping system with a stainless hose clamp, and then run the wire to the ground rod.

ditions, where the pump speed will always match the output of the solar thermal collectors. On the other side of the coin, at the end of the day, when the solar thermal collectors are cooling off and the storage tank is hot, the pump is still circulating, which could result in a slight cooling of the storage tank. During low demand conditions mentioned above, the photovoltaic-powered system will continue to circulate the solar fluid, which helps extend the fluid's life by keeping it from overheating.

In the end, the advantages and disadvantages of each system from a thermal performance standpoint are a wash. The most reliable type of system is a photovoltaic-powered system because it does not require a controller, which is probably the most vulnerable component in a solar water heating system.

ADJUSTING A SOLAR RADIANT FLOOR HEATING SYSTEM

In most cases, only seasonal adjustments are necessary on your solar radiant floor heating system. You have a ball valve located on the liquid-to-liquid heat exchanger bypass. That valve should remain around one-third closed all the time. This valve forces some of the hot solar fluid coming down from the collectors through the heat exchanger while allowing most of the fluid to go directly to the radiant floor or to the shunt loop, depending on the season. You need enough flow through the heat exchanger to get enough hot water, while still getting a significant flow to the heating system. Observe the temperature of the tank after a sunny day. If it is above or below 140°F, then adjust the valve accordingly.

In the spring, when no more heating is required in the home, open the shunt loop valve and close the valve to the radiant floor. Reverse this procedure in late summer or early fall, depending on your heating requirements. You would like to start charging the floor as early as possible in the fall. You can adjust the temperature in the home by keeping windows open. Each individual home will react differently, and no two years are exactly alike regarding weather patterns. Of course you don't want to cook yourself out of the house during late summer, so experiment.

Maintenance

Like any other mechanical devices or products, solar energy systems require periodic maintenance. The good news is that a properly designed and installed system will require very little. You can further ensure minimum maintenance by choosing high quality parts when installing your system in the first place. I have seen people choose inferior products for their systems because they are cheaper, while others choose only high quality components. The overall price difference between these two systems may be only several hundred dollars, which is insignificant to the overall system cost. Over the long run, the system ends up costing a lot more because of service calls on those inferior components' failures. I suggest you save yourself money and headaches down the road by choosing high quality products in the first place.

PERIODIC INSPECTION

It is wise to periodically look over your solar energy system to make sure it is working and that there are no glaring problems. Once a month you should look at your collector array and observe if there is anything out of the ordinary with it. Glance at the collectors to make sure they are in one piece. Notice any discoloration or dripping. Look for any loose pipe insulation or hanging wires. You can also check once a month to make sure the system is collecting heat. After a sunny day you should check the solar storage tank. If it is hot, all is well. This should take about two minutes a month. Note that it is normal to see condensation on the inside of the glass of flat plate collectors of all kinds. Flat plate collectors must breathe and they blow out air when they get hot and suck in air as they cool. If the air that they suck in has a lot of moisture in it, that moisture ends up condensing on the inside of the glass. This amount of water inside of a collector does not deteriorate it.

In addition to periodic visual inspection of your system, some periodic maintenance may be required that is specific to your type of system. I have itemized the main topics below. As there are only three kinds of solar water heating systems that I recommend, I am only detailing issues for those systems.

DIFFERENTIAL TEMPERATURE CONTROLLERS

Because differential temperature controllers are used in several types of systems, I have included a special section just about troubleshooting them. To perform some of the tests listed below, you will need a digital multi-meter.

If your system is not operating, the first thing to check is the controller. All controllers will have at least two lights on the front, one indicating that power is being supplied to the unit and the other indicating whether the pump is turned on or off. If both lights are on, take your multimeter and see if power is going to the pumps. If there is power going to the pumps, your problem is with the pumps. If both lights are off, first check that there is power at the unit by using a multi-meter. If there is power at the controller and no lights are on, the controller has failed and should be replaced.

Sometimes the power indicator light will be on but the controller is not turning the pumps on at the proper times. This could be caused by a sensor failure, corrosion at sensor wire connections, or a controller failure. First, visually inspect the circuit board of the controller. If there are any black blotches on the circuit board, it is probably shot. This is often observed after a lightning storm. If there are no obvious burn marks on the circuit board, you can test the controller by removing the sensor wires one pair at a time. First remove the collector wires from the controller terminals and short-circuit the collector sensor terminals with a jumper wire,

which makes the controller think that the collector is hot. The pumps should come on. If the pumps come on when you short-circuit these terminals, there is a problem with the collector sensor or the sensor wire connections. You can also short-circuit the storage sensor terminals, which makes the controller think the tank is very cold. It should turn the system on using the same procedure. Remember that either a sensor or a jumper must be attached to both pairs of sensor terminals in order to perform this test. If the controller checks out the next thing to check is the sensors.

First inspect all connections on the sensor wire circuits for corrosion. These connections should be soldered. If these connections look good, you should then check the sensors. To check a sensor, disconnect the sensor wire from the controller and attach the sensor wires to your digital multimeter and set the meter to measure ohms resistance. If you get no resistance or infinite resistance, the sensor has failed and needs replacing. Most controllers have a sensor resistance chart included with the instructions, so if you have a reading, you can actually tell the temperature of the sensor by comparing the meter reading with the chart. If the reading gives a temperature that is obviously wrong (like the sensor saying it is 500°F), the sensor is probably bad and needs replacing. If you do not have a sensor chart, a 10K sensor will have a resistance of 10,000 ohms at 77°F and a resistance of 2,043 ohms at 150°F. You can see that as the temperature rises, the resistance goes down.

If the sensors check out and the controller still does not do its thing, replace the controller.

ICS SYSTEMS

If you live in a climate that never experiences freezing conditions, your ICS system will be in operation all year. The only periodic maintenance required is monthly visual inspections.

If your ICS system is a seasonal system, you will be draining your system at the end of the season and filling it at the start of the season. If the collector is installed in a climate where freezing conditions occur, it is critically important to get the system drained well before freezing conditions are expected. I suggest that you do not push your luck by waiting till the last moment. When draining a system it is most important to get every bit of water out of the pipes. Even a couple of cupfuls of water can

burst a pipe if it freezes. Proper pipe layout and collector orientation during installation will facilitate efficient drainage. Leave all drains open while the system sits idle. If you are unsure that the system has drained properly, use compressed air to blow out the lines. A few minutes of diligence at this time can save big headaches and money the following spring.

DRAINBACK SYSTEMS

Here is what can go wrong with a drainback system: the main pump can fail and the system will simply not work; there could be a loss of solar fluid and the system will simply not work; the controller could fail and the system will simply not work; and the system could freeze and serious complications will arise.

You can detect the first three potential problems by visual inspection. If the pumps fail, you will notice that the system is not heating the solar storage tank. If either of the pumps fails, they fail quickly. Either they work or they don't. If the main circulator pump fails, there will be no heat in either the drainback tank or the solar storage tank after a sunny day. If the heat exchanger pump fails (if one is in the system), then the drainback tank will be hot but the solar storage tank will be cold during operation. If the controller fails, you will notice that neither of the pumps will be running when it is sunny out and the solar storage tank will be cold. Note that if the solar storage tank reaches its high limit, the system may be off even if it is sunny outside, so make sure to check the storage tank for heat if the pumps are not running and it is sunny.

All drainback tanks should be fitted with a sight glass. This sight glass is mounted on the side of the drainback tank and it shows the heat transfer fluid level within the tank. This fluid level should be checked periodically. It is normal for some of this fluid to evaporate over time, so annual close inspection should be done and fluid should be added when required.

System freezes can be caused by a controller failure or climatic conditions. If a controller does not shut the system down properly, the distilled water that is used as the solar fluid could freeze in the collector or in the collector piping. If a freeze occurs, make sure to check the controller for proper operation before putting the system back online.

CLOSED-LOOP ANTIFREEZE SYSTEMS

The wear parts of this type of system are the pumps, the solar fluid, and the expansion tank(s). All these components should last between 10 and 20 years before they need replacement. Your monthly inspections should identify if the system is working or not. If the system is working, the only thing to do is to verify that the system is holding its pressure. Starting with year ten, the antifreeze should be checked in the fall of the year before serious freezing conditions would be expected. Checking the antifreeze is outlined below. If the antifreeze readings show the fluid is in near-new condition, schedule the next check for five years. If the readings show degradation but still fall within the acceptable range, schedule the next check accordingly. When the antifreeze check shows that the fluid should be replaced, you should replace it. At this time I also like to replace the expansion tank(s).

In a closed-loop antifreeze system, if the system is not circulating and the controller and pumps check out okay and the system is still under pressure, the problem could be caused by a failed expansion tank. Expansion tanks that have bladders eventually wear out from the bladder's constant flexing. I always choose an expansion tank that has an extra-heavy-duty bladder. A good bladder-type expansion tank should last 15 to 20 years or more. Bladder expansion tanks should always be mounted below the pipe they are attached to. This helps prolong the life of the bladder by keeping it much cooler than if it were above the pipe it is attached to. You can test a bladder expansion tank by tapping the bottom of the tank. It should sound hollow. If the tank does not sound hollow, release a very small amount of pressure from the Schrader valve on the bottom of the tank. If air comes out, the tank is probably OK. If liquid comes out, the tank is broken and should be replaced.

CHECKING ANTIFREEZE

The proper kind of antifreeze to be used in these systems is propylene glycol. As this fluid breaks down, it loses its ability to protect the system from freezing, and becomes more acidic. Special testers are needed to check for these conditions. Regular automotive testers will not work. The most common kind of tester is optical. You put a drop of antifreeze on the appropriate surface of the tester, point the tester at a light source, and look into the instrument. You will see a scale that indicates the freeze protection. These meters vary between manufacturers, so follow the directions that come with the tester. Make sure you purchase a tester that measures propylene glycol antifreeze. You can get these meters from some auto supply stores, from some solar system manufacturers, or from a scientific instrument company. You should also test the pH of the antifreeze. You use a pH meter for this test or use litmus paper. The pH should never be below 7.5. Note that the smaller the number, the higher the acid content, and 7 is neutral.

SOLAR SWIMMING POOL SYSTEMS

At the start of the heating season you should check the controller for proper operation. Visually inspect the collectors and piping. Make sure all the fittings are still connected. Clean any debris that may have accumulated around the collectors. Turn the system on and check for leaks. Turn the system off and make sure it drains properly.

AIR COLLECTOR SYSTEMS

The weak links in air systems are the dampers. Even the best of them just do not last very long. The dampers should be checked and lubricated at least once a year at the start of the heating season. The blower should also be checked and lubricated at the start of every heating season.

One thing that can impact the performance of air collectors is dust. On all air systems except transpired systems, air is drawn from inside the building and circulated through the collector. If this air has any dust in it, that dust can build up on the absorber plate and significantly impede heat transfer and decrease the collector's efficiency. To minimize this, an air filter must be placed in the cold air duct. This air filter must be checked periodically, especially where dusty conditions exist. Media filters with lots of surface area are the best filter choice and will last the longest between replacements.

For additional current information about solar thermal systems, please refer to Tom Lane's book, *Lessons Learned 1977 to Today*, available from the author at www.ecs-solar.com

Appendix A

WIRE SIZING CHART

Amps	#14	#12	#10	#8	WIRE SIZE #6	#4	#2	#1/0	#2/0	#4/0
1	45	70	115	180	290	456	720			
2	22.5	35	57.5	90	145	228	360	580	720	1060
4	10	17.5	27.5	45	72.5	114	180	290	360	580
6	7.5	12	17.5	30	47.5	75	120	193	243	380
8	5.5	8.5	15	22.5	35.5	57	90	145	180	290
10	4.5	7	12	18	28.5	45.5	72.5	115	145	230
15	3	4.5	7	12	19	30	48	76.5	96	150
20	2	3.5	5.5	9	14.5	22.5	36	57.5	72.5	116
25	1.8	2.8	4.5	7	11.5	18	29	46	58	92
30	1.5	2.4	3.5	6	9.5	15	24	38.5	48.5	77
40			2.8	4.5	7	11.5	18	29	36	56
50			2.3	3.6	5.5	9	14.5	23	29	46
100					2.9	4.6	7.2	11.5	14.5	23
150							4.8	7.7	9.7	15
200							3.6	5.8	7.3	11

Maximum Distance (feet)

12 Volt 2% Wire Loss Chart

223

Appendix B

METRIC CONVERSIONS

To Convert	From Metric	To US	Multiply By
Temperature	C	F	9/5 + 32
Delta Temperature	C	F	1.8
Length	meters	feet	3.28
Length	millimeters	inches	0.04
Area	square meters	square feet	10.76
Volume	cubic meters	cubic feet	35.31
Liquid Volume	liters	gallons	0.26
Storage volume/	liters/	gallons/	
collector area	square meters	square feet	0.024
Speed	meters/second	mph	2.2369
Power	watts	btu/hour	3.41
Power	watts/square meter	watts/square foot	0.09
Pump power	watts	horsepower	0.0013
Pump power	kilowatts	horsepower	1.34

To Convert	From Metric	To US	Multiply By
Heat exchanger power	kilowatts	btu/hour	3412
Solar resource	kWh/m^2/day	btu/ft^2/day	316.99
Solar radiation	MWh/m^2	Million btu/ft^2	0.32
Energy	MWh	Million btu	3.41
Energy	liter	gallon	0.26
Energy cost	$/MWh	$/Million btu	0.29
Energy cost	$/liter	$/gallon	3.78
Energy cost	$/cubic meter	$/ccf	2.83

(ccf=100 ft^3=1 therm)

GLOSSARY

Absorber – The part of a solar collector that absorbs the sun's energy and changes that energy into heat.

Absorptance – The efficiency of an absorber. A ratio of the amount of energy hitting a surface to the amount of energy that the surface actually absorbs.

Absorptive coating – A coating for absorber plates that maximizes the plate's absorptance.

Active solar energy system – Any solar energy system that uses a collector to absorb solar energy.

Ambient – The temperature of the surrounding air.

Aquastat – A thermostat that measures the temperature of a liquid.

Azimuth – The distance in degrees from north. (South would have an azimuth of 180).

Ball valve – A valve that uses a ball with a hole through its center to control the flow of liquid through a piping circuit.

Batch heater – Another name for ICS collector.

Battery – A storage device for energy.

Boiler control – A hydronic heating system component that includes a low voltage power supply and a heavy-duty relay for switching purposes.

British Thermal Unit (Btu) – The amount of energy it takes to raise the temperature of one pound of water one degree F.

Centrifugal pump – A pump that moves a fluid by spinning it with enough force to throw the fluid outward.

Check valve – A valve that allows a fluid to travel in only one direction within a circuit.

Circulator – A pump designed to move a fluid within a circuit.

Closed-loop solar heating system – A type of system where the solar fluid stays in the solar loop and does not mix with the domestic water.

Collector – A device that collects solar energy.

Collector loop – Another name for solar loop.

Collector tilt – The angle between the solar collector plane and the horizontal plane.

Conduction – Heat flow within a material that is caused by the difference of temperature within that material.

Convection – The movement of parts of a fluid because of variations in the fluid's density caused by temperature differences.

Counter-flow heat exchanger – A heat exchanger where the two fluids pass each other in opposite directions.

Coupler – A pipe fitting that joins two pieces of pipe together.

Degree-day – The difference between 65°F and the ambient temperature.

DHW – Domestic hot water.

Differential temperature controller – An electrical device that measures the difference in temperature between two locations and switches a pump on or off in relation to the difference in the temperatures at the two locations.

Diffuse radiation – Solar radiation that is scattered by the atmosphere and everything that is suspended in the atmosphere, particularly water vapor.

Direct radiation – Solar radiation that has not been scattered.

Double-walled heat exchanger – A liquid-to-liquid heat exchanger whose separation wall between the two fluids is made of two layers.

Drainback – A solar water heating method where the solar fluid is pumped to the collectors and fills the solar loop piping when solar energy is available for harvest but drains back to a holding tank when not collecting energy.

Draindown system – A solar water heating system that drains when not in use.

Emittance – The property of a material to radiate energy.

Expansion tank – A tank used in solar energy systems to compensate for the expansion of liquids in a closed-loop circuit.

Evacuated tube collector – A collector that uses absorber plates that are enclosed in a glass tube that has a vacuum inside of it.

Flat plate collector – A rectangular solar thermal collector, typically four inches deep, four feet wide and eight feet or ten feet tall.

Flow meter – A device that measures the flow of a liquid within a pipe. It is often a clear pipe with a scale on the side.

Flow rate – A measure of the rate of a certain volume of fluid flowing through a circuit.

Fossil Fuel – A carbon-rich and energy-dense fuel that was created from concentrated organic matter that lived or grew on the earth millions of years ago.

Friction – The resistance to movement created when two materials rub against each other.

Friction head – The pressure created in a circuit resulting from a fluid moving through a pipe.

Glazing – A transparent and weatherproof covering.

Head – The total pressure or resistance to the flow of a liquid within a circuit exerted by gravity and friction.

Header – A manifold.

Heat exchanger – A device that facilitates the transfer of heat from one fluid to another.

Heat transfer fluid – A fluid that is used to transfer heat from one location to another. This fluid could be a liquid or a gas.

Hydronic – Hydronic heating systems use a liquid to transfer heat. Forced air systems use air.

Indirect system – A solar water heating system where the domestic water is not heated in the collectors but is heated using a heat transfer fluid and a heat exchanger.

Impellers – The vanes in a pump that spin and move the fluid.

Insolation – The total amount of solar radiation hitting a surface.

Liquid collector – A collector that uses a liquid as the heat transfer fluid.

Manifold – A pipe with several outlets.

Open-loop system – A kind of solar water heating system where some part of the system is vented, or the solar loop contains potable water.

Orientation – The direction the solar collectors face in relation to South.

Passive systems – Solar heating systems that do not use a collector to absorb the sun's energy. A term used to describe a solar heating system where the house itself is the solar collector.

Photovoltaic panel – A type of solar collector that absorbs solar energy and converts it to electrical energy. Also called PV.

Potable water – Drinkable water.

Propylene glycol – A non-toxic antifreeze used in solar heating systems. It is mixed with water and used as a heat transfer fluid.

Pump – A device that moves a fluid through a circuit.

Pump curve – A chart that shows the performance of a pump over varying conditions.

Radiation – The flow of energy through a space.

Renewable energy – Energy that is replenished by nature on a regular basis.

Risers – The pipes that connect the top and bottom manifolds in an absorber plate.

Selective surface – A surface that absorbs solar energy very efficiently but does not radiate solar energy very well.

Sensor – A device that identifies the temperature at a certain location.

Solar collector – A device that collects the energy of the sun that is radiated to the earth's surface.

Solar energy – Energy produced by the sun that is radiated to the earth's surface.

Solar fluid – The heat transfer fluid used in a solar heating system.

Solar loop – The circuit of piping that travels from the collectors to the heat exchanger/storage and back to the collectors.

Specific heat – The amount of heat, in Btu, needed to raise the temperature of one pound of a substance by one degree F.

Stagnation – When the solar fluid in a closed-loop solar heating system is not circulating.

Static head – The pressure created by a column of water in a pipe.

Therm – 100,000 Btu.

Thermosiphon – The movement of a fluid caused by convection.

Union – A pipe fitting that joins two pipes together and is capable of coming apart.

Volute – The part of a pump that houses the impellers. Also called a pump head.

Wrench – A solar installer.

INDEX

Bold numbers indicate illustrations

A

absorber plates, 24–25

acid rain, 10

active systems, 60

actuators, 88–89

aesthetics, 22

air-based heating systems: and air quality, 118–119; description, 118–122; maintenance, 222; mounting, 119–120, 175; sizing, 149–150

air collectors, **33**, 121

air conditioning systems, 110

air heaters, 33

air locks, 206

air pressure, 46–47, 55–56, 57–58, 81. *see also* pressure gauges

air vents, **52**–53, 84–85

anode rods, 36

antifreeze, 44–46, 72, 78, 220, 221. *see also* closed–loop antifreeze systems

aquastats, **52**, 93, 98, 108, 109, 120

Armaflex, 43, 189

arrays: mounting ground, 172–175, **173–174**; mounting roof, 165–172, **169–170**; saw-tooth, **171**; space heating, 96, 98, 143–144; swimming pool, **88**, 89; water heating, 136–139

automatic fill valves, 85

B

Bailey, William J., 3

ball valves, 54–**55**, **81**, **84**, 85, 117

baseboard units, 111

blowers, 108, 109, 110, 111

boiler drains, **178**

brackets, mounting, 160–161

bulkhead fittings, 47, 105

C

Carter, Jimmy, 4

cartridge circulators, 41

catwalks, 171–172

charging a system, 190, 203–209, 206

check valves, 49–51, **50**, 57, 80–81

climate: and closed-loop antifreeze systems, 76; and Dole valves, 53–54; and drainback systems, 70–71, 72, 100; and evacuated tube collectors, 27, 29–30; and flat plate collectors, 23, 25, 29–30; and flooded open-loop systems, 68–69; and high-mass systems, 115, 216; and ICS collectors, 21, 23, 63, 64; and maintenance checks, 220; and pipe insulation, 43–44, 189; and snow on collectors, 27, 171–172; and solar fluids, 44–45, 46; and space heating systems, 96, 98,

233

ABOUT THE AUTHORS

Bob Ramlow

Benjamin Nusz

Bob started in the renewable energy business in 1976 and has been involved in sales, design, installation and manufacturing. He has taught workshops about solar across the U.S. and is a founder of the Midwest Renewable Energy Association. He continues to do solar consulting from his home in Wisconsin.

Benjamin currently works as a solar water heating consultant and site assessor in Wisconsin.

If you have enjoyed *Solar Water Heating*, you might also enjoy other

BOOKS TO BUILD A NEW SOCIETY

Our books provide positive solutions for people who want to make a difference. We specialize in:

Sustainable Living • Ecological Design and Planning • Natural Building & Appropriate Technology
New Forestry • Environment and Justice • Conscientious Commerce • Progressive Leadership
Educational and Parenting Resources • Resistance and Community • Nonviolence

For a full list of NSP's titles, please call 1-800-567-6772 or check out our web site at:

www.newsociety.com

New Society Publishers

ENVIRONMENTAL BENEFITS STATEMENT

New Society Publishers has chosen to produce this book on recycled paper made with 100% post consumer waste, processed chlorine free, and old growth free.

For every 5,000 books printed, New Society saves the following resources:[1]

35	Trees
3,160	Pounds of Solid Waste
3,477	Gallons of Water
4,536	Kilowatt Hours of Electricity
5,745	Pounds of Greenhouse Gases
25	Pounds of HAPs, VOCs, and AOX Combined
9	Cubic Yards of Landfill Space

[1]Environmental benefits are calculated based on research done by the Environmental Defense Fund and other members of the Paper Task Force who study the environmental impacts of the paper industry.

NEW SOCIETY PUBLISHERS